ROCKET BELT
Pilot's
Manual

A GUIDE BY THE BELL TEST PILOT

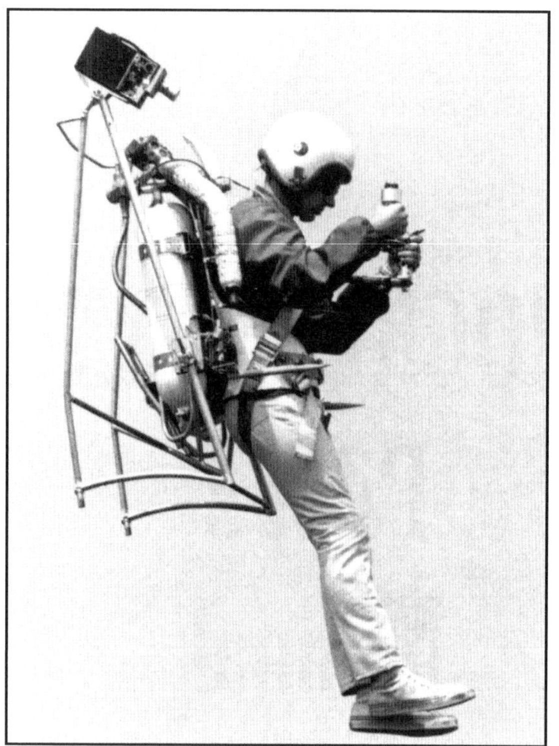

Limit of Liability / Disclaimer of Warranty
The author and publisher of this work have used their best efforts in preparing this manual. The publisher and the author make no representation. or warranties with respect to the accuracy or completeness of the contents of this manual and specifically disclaim any implied warranties of merchantability or fitness for any particular purpose and shall in no event be liable for any loss of profit or any other commercial damage, including but not limited to special, incidental, consequential, and other damages.

We acknowledge the financial support of the Government of Canada through the Book Publishing Industry Development Program for our publishing activities.

Published by Apogee Books an imprint of Collector's Guide Publishing Inc., Box 62034, Burlington, Ontario, Canada, L7R 4K2, http://www.apogeebooks.com

Printed and bound in Canada

The Rocket Belt Pilot's Manual by William P. Suitor
ISBN 9781-926592-05-3 - ISSN 1496-6921

Cover: Robert Godwin

The Rocket Belt Pilot's Manual

A Guide By The Bell Test Pilot

by

William P. Suitor

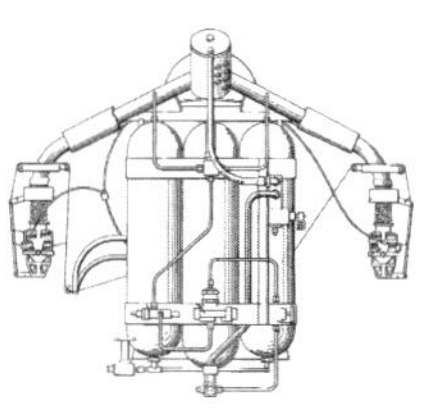

An Apogee Books Publication

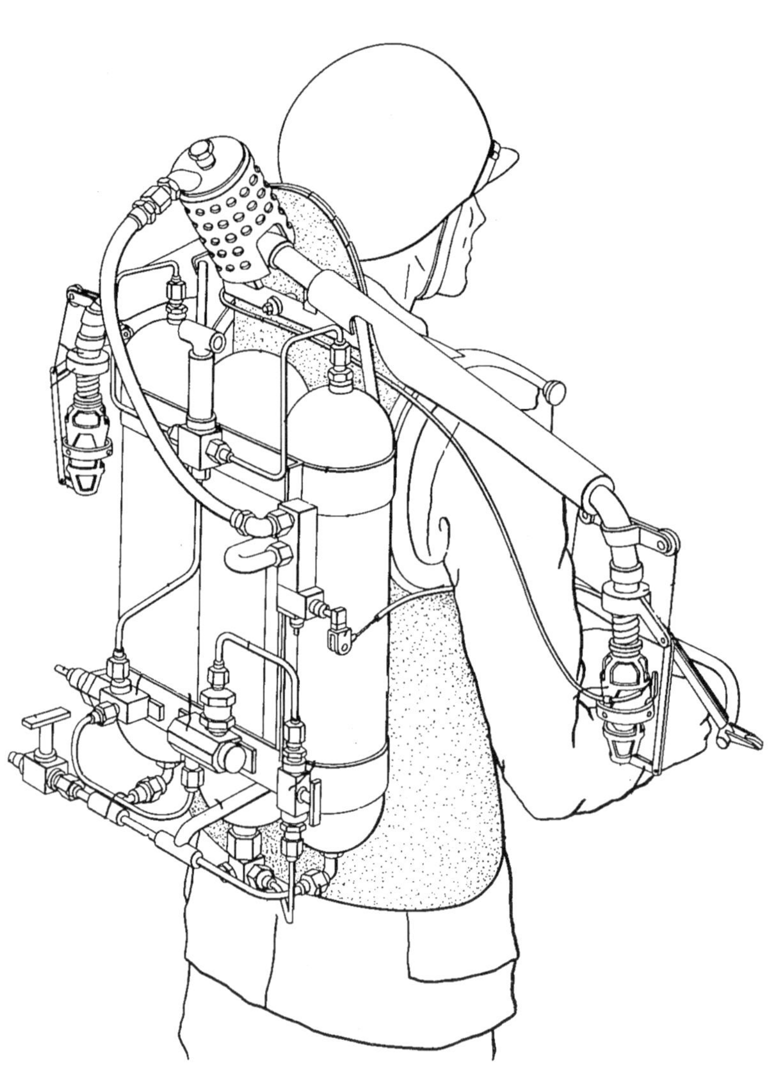

Table of Contents

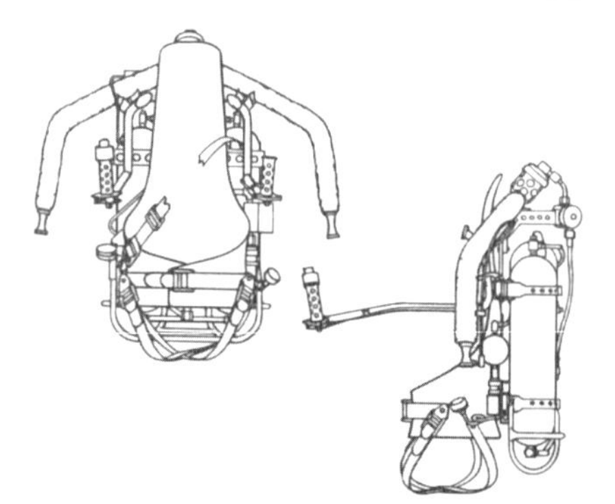

ROCKET BELT

On April 20, 1961 a "Small Rocket Lift Device" was successfully flight tested. This unique system marked the beginning of a totally new dimension in the realm of flight.

This device, known as the Bell Rocket Belt, was the first rocket powered system in the world which propelled man above the ground in controlled free flight.

The Bell Rocket Belt is a hydrogen peroxide propulsion system mounted on a fiberglass corset.

Lift is provided by thrust from twin rocket nozzles which are fed by a central gas generator controlled by a throttle. Metal control tubes extend forward on each side of the operator. A control handle on one tube permits the operator to change his flight direction. A motorcycle-type hand throttle on the other tube allows him to regulate rocket thrust levels, thus controlling his rate of climb and descent. These controls permit complete freedom of flight, including forward, backward, sideward, up and down movement and even permits hovering in mid-air.

Since the Rocket Belt was first unveiled to the public at Fort Eustis, Virginia, in June 1961, more than 3,000 demonstrations have been conducted throughout the United States and Canada and on five continents.

The Rocket Belt concept presented by Textron's Bell Aerosystems has fostered new challenges for earth and space transportation. Present research and development efforts have successfully flight tested three new rocket-propelled systems for transporting men and equipment on earth or over the surface of the moon. The new systems, extensions of the Rocket Belt concept, are a small Flying Chair and a standup version for transporting one or two men.

Although the present earth bound capabilities of the Bell Rocket Belt are limited to feasibility studies and public demonstrations, the freedom of movement and the restart properties of the Rocket Belt are an important feature for future lunar transportation applications.

SPECIFICATIONS

PROPELLANT	Hydrogen Peroxide (H_2O_2)
PROPELLANT WEIGHT	47 pounds
EMPTY WEIGHT	63 pounds
THROTTLEABLE THRUST	0-300 pounds
MAXIMUM RANGE	866 feet
ALTITUDE	80+ feet
MAXIMUM SPEED	60+ miles per hour
RECORD OF RELIABILITY:	100% demonstrated reliability in more than 3,000 flights

04-8J8

ACKNOWLEDGEMENTS

There were many people who gave me assistance in writing this book, too many to thank each individually. Even if I tried, I am sure I would inadvertently miss a few. Nevertheless, to all of them, a sincere thanks for their support and assistance.

I must single out a few who were the backbone of my research effort: Marve Goldpenny and Stan Smolin of Bell Aerospace, Bob Roach of Calspan, and Mr. Chuck Kreiner, the Gentleman's gentleman who used to head the Bell Rocket Belt Demonstration Team. A special thanks must also go to Dr. Colyn Durrery of the State University of New York at Buffalo for the use of the Bell B-4 Rocket Belt for my photos. And, finally, to my family and friends for putting up with my constant mumbling about my work.

In early 1997 as a result of my friendship with Doug Malewicki the famed propulsion genius, I was introduced to Chris Wells of Boulder, Colorado. Chris shares an interest in Wendell's dream and felt it was time I got serious about this publishing endeavor. So, after almost ten years of my wandering about the publishing world, Chris took the bull by the horns, edited my manuscript, scanned in all my photos, and essentially transformed that collection of thoughts and knowledge I called my book into this truly interesting and readable work of literary art. Probably more than anyone, Chris is responsible for you the reader being able to enjoy the book. To Chris I owe a sincere debt of gratitude and thanks for believing in me and my efforts.

Then for ten more years I fiddled around and still did not publish it. Enter Andrew Filo and the 2007 Rocket belt Convention in Niagara falls.I would have to say that Andy Filo is one of the worlds leading rocket belt enthusiasts, as well as having an extremely creative mind. Andy hauled me and my manuscript and archives to his home in Silicon Valley and for a week we pored over what I had...and, we found a publisher ! Now in late 2009 thanks to Andy's generosity and encouragement, I am a "Published Author" !

I would be remiss to not mention my late, great friend from Bell, Dan Paul and "The Girl From Herkimer"...Dan, I finally did it !

I wish to dedicate this book to Wendell F. Moore, the "father" of the Rocket Belt. Also, to Nelson Tyler, who kept "The Dream" alive.

A note of gratitude to my late, great, friend Joe Wright, who met a senseless and untimely death in July 1998: "Joe, wherever you are, we've done it: the book is in print!"

Salute!
Bill Suitor

Note to the Reader

Although I use the words he, his, or him to refer to unknown people, this is only because it seems easier to read than always saying he or she.

Unfortunately, there are not currently any formal training programs for flying a Rocket Belt. The things I describe in this manual were part of a training program in the 1960s.

Watch for Bill's two other upcoming books: "Rocketman", and "Beyond Rocketman".

Some of the material in Chapters 3 and 7 was taken from:
"V/STOL Concepts and Developed Aircraft, Volume 1: A Historical Report (1940-1986)", by Bernard Lindenbaum, Universal Energy Systems, November 1986, under sponsorship by the Air Force Wright Aeronautical Laboratories. Report # AFWAL-TR-86-3071; NTIS Report # AD-A175-379.

This is a wonderful technical reference for those interested in rocket powered and turbojet flying platforms.

CHAPTER 1 Introduction

Icarus never did...

Sky Sentry never did...

So you want to fly a Rocket Belt? Good! It has been one of man's biggest dreams since he first saw birds in the air. The power of free flight, wingless, rotorless, no strings. Your body is the fuselage; your legs are the landing gear. FLIGHT, the ability to just think about it and then, as easily and naturally as riding a bicycle, you fly, pure, simple, unencumbered ... FLIGHT!

A company called Bell Aerosystems built the original (and most successful) Rocket Belt. I have been a Rocket Belt pilot since 1964; I started at Bell, and went on to fly all but two of the rocket belts ever successfully flown to date. People often ask many questions about how it felt, how I was trained and how the Belt worked. I will try to teach you to fly the Belt the same way I learned at Bell Aerospace. The training was always a matter of serious, safety-first instruction, but was also exciting, interesting, and enjoyable.

But Bill Suitor did, at age 19!

The Bell Rocket Belts compiled an unbelievable 100% safety record in over 3,000 flights from 1961 to 1969. This fantastic record was due to the unerring professionalism of the entire Bell team, especially the technicians. A special debt of gratitude is owed to Ernst "Ernie" Kreutinger, the Bell crew chief, for his strict attention and adherence to detail. His motto (that everyone on the project had to live and work by) was "We will not deviate"; the safety record proves his genius.

So, with this motto, I will begin..

You must remember at all times that this is not some simple contraption or circus act, but a very powerful and dangerous piece of space-age technology. Construction should only be attempted by highly-skilled members of aerospace research teams, thoroughly versed in handling explosive fuels, and with extensive backgrounds in metallurgy to know exactly what types of materials to use.

,,,thanks to Ernie Kreutinger, shown here with the Bell "A-1" Belt.

There are several basics to flying a Rocket Belt that you must never forget. First, without the engine running, the Rocket Belt flies like an anvil. Second, any landing you can walk away from is a good one. Third, never land at any speed faster than you can run. As for glide slope and rate of descent, never exceed the maximum impact strength of your sneakers!

CHAPTER 2

A Brief History of the Rocket Belt

Impetus for the flying belt concept came from Colonel Charles Parkin of the U. S. Army Transportation Research and Development Command (TRECOM). His personal interest in the idea of using rocket power to improve a soldier's physical capabilities began in the 1940s.

The Bell Aerosystems Company (now Bell Aerospace) of Buffalo, New York, successfully developed the world's first Rocket Belt. The key man behind this effort was Wendell F. Moore, a rocket propulsion engineer. Moore conceived of the idea in 1953 while working on the Bell X-l project, actually making, his first designs in the sands at Edwards Air force Base for his friends and co-workers. He finally applied for a patent on his design in June 1960, for which U.S. Patent No. 3,021,095 was granted on February 13, 1962.

Bell began funding exploratory work, with Moore in charge, in 1957 after learning of the Army's interest in individual mobility. To prove his idea, a tether system was used with a high-pressure nitrogen hose propelling the rig during the first attempts. Test flights, piloted by Moore and two others, commenced on December 17, 1957. In spite of its crudeness, the nitrogen gas rig proved the feasibility of underarm lifting, and that stable operation was possible using kinesthetic control.

Wendell F. Moore

Kinesthetic control is the use of body movement to provide attitude control by shifting the location of the center of gravity (CG) with respect to the lift force. The degree of success depends on each individual pilot's

Wendell on nitrogen gas hose, 1957

instinctive reactions. One man might be able to control his flight easily, while another cannot.

In 1959, TRECOM issued a formal request-for-proposal to industry to determine the feasibility of building a Small Rocket Lift Device (SRLD) to increase a soldier's mobility. So, in 1960, the Bell team built a Rocket Belt. It used hydrogen peroxide (H_2O_2) for fuel.

The Army contract sought to establish concept feasibility at minimum cost, so proven off-the-shelf equipment was used whenever possible. On December 29, 1960, Moore made the first tethered flight. Although Moore was the first man to attempt to fly it on the tether, he was not the first person to solo successfully. He was injured in a freak accident during testing and had to discontinue training.

The honor of the first untethered flight belongs to Harold Graham, another Bell rocket engineer. At 7:30 a.m. on April 20, 1961, Graham made aviation history by becoming, the first human in history to fly in controlled free flight using a rocket attached directly to his body. Look out Buck Rogers!

His first flight lasted just 13 seconds and covered only 112 feet, but history had been made.

With Bell's successful demonstration of SRLD, the contract was terminated in May 1961. The company continued Rocket Belt development efforts on its

Harold Graham

own; between 1961 and 1969 Bell's pilots made more than 3,000 flights with 5 Rocket Belts. Significantly, the H_2O_2 fuel system had been 100% reliable.

The final version of the Bell Rocket Belt had the following specifications:

Thrust	0-330 pounds
Empty Weight	63 pounds
Propellant Weight	47 pounds
Takeoff Weight	110 pounds
Flight Duration	21 seconds

... and the following "officially measured" records had been set by 1964:

Maximum Distance	866 feet
Maximum Altitude	80 feet
Maximum Speed	60 mph

I rarely attempted flights longer than 200 yards. I set that 866-foot distance record on only my 15th free flight, when I was 19 years old. Young, inexperienced, and running on luck! By the time of my flight at the 1982 World's Fair in Knoxville, Tennessee, I know I had flown over 1,200 feet horizontally and reached 140 feet in altitude. Also, I probably exceeded 80 mph in a 1966 over-water flight only out of desperation to make it back to shore !

With Ernie

Harold flying the "A-1" belt, 1961

Jetbelt in flight

Aircab

CHAPTER 3 

Other Rocket Belt Developments

Wendell in "Aircab" prototype in 1968

The success of the Rocket Belt and the desire for longer flight duration and range led to several further developments, financed by company funds. Moore designed a fuel pump system to eliminate the nitrogen tank, permitting a larger fuel supply. He also considered the addition of wings, making a patent application for this concept in 1962.

In June 1962, Bell propulsion engineer John K. Hulbert proposed a rocket-driven shrouded-fan system for the Rocket Belt. Although a test model was built, work on this concept was discontinued in favor of a Jet Belt. The Jet Belt made flights of up to 5 minutes, but the military had lost interest.

Starting in 1965, Bell used the Rocket Belt as the basis for a one-man stand-on platform device called the

"Aircab" flying ejection seat.

"Pogo." Under contract to NASA during 1966 to investigate potential lunar transportation devices, the Rocket Belt developed into the Rocket "Pogostick", seated vehicles, and a two-man Pogo. These were used to investigate handling qualities and control methods for lunar and Earth mobility systems.

An interesting aside of the SRLD program was the trainer created by Bell to reduce expense and increase safety during pilot training. The equations of motion for the man-machine combination were built into an analog computer. The pilot trainee learned to handle the Rocket Belt controls by observing a moving image displayed on an oscilloscope screen. This type of trainer was far ahead of its time.

Hulbert's fan system

The five Bell Rocket Belts are no longer flying. The Army Transportation Museum (Fort Eustis, Virginia); the Buffalo Museum of Science (Buffalo, New York), and the University of Buffalo (Buffalo, New York) each have one on display. The Smithsonian Museum (Washington,

Bell Jet and Rocket belts

DC) has another on display. One of the Belts was dismantled so the parts could be used for the Flying Chair and the Pogo.

There has been a continued interest in the Rocket Belt idea. Several more Belts have been constructed over the years. All of these attempts were just duplicates of Wendell Moore's original work; no improvements could be made to his original design. All are limited to under 35 seconds of flight time, and used the same dangerous 90% H_2O_2 fuel.

In 1969, Nelson Tyler of Los Angeles, California, built a copy of the Bell Rocket Belt, called the NT-1. It was flown in all sorts of movies, commercials and exhibitions. I flew NT-1 in the Opening Ceremonies of the 1984 Olympics. An amusement park called Tivoli Gardens in Denmark purchased the Belt, and it has since fallen into disrepair and no longer flies.

Then, in the early 1990s, Kinnie Gibson built his copy of the NT-1 and currently employs one demo pilot. Kinnie owns a company called Powerhouse Productions in Texas.

By the early 1990's the Tyler Belt, NT -1, was getting long in the tooth and needed replacing, Gibson set about that task with his partner Brad Barker. A disagreement between the two of them led to Barker joining with two other partners, Joe Wright and Larry Stanley, to build the RB-2000 rocket belt that I flew in 1995. I named it "Pretty Bird". Gibson continued with his new belt.....that looks VERY similar to "Pretty Bird" only it's green rather than red....and Eric Scott began flying it for Gibson.

"Pretty Bird" RB-2000 is a copy of

POGO

Tyler's NT-1, as is the Gibson version. They tried to extend the flight time by using lighter structural materials so more fuel could be carried. Unfortunately, RB-2000 was grounded after only a few flights, due to business and technical problems, and now sits in storage somewhere.

By 1995, they had both added larger fuel tanks which extended the flight time to roughly 30 seconds.....whoopie!

Enter Troy Widgery, founder of "Go Fast Sports", he hired Eric Scott away from Gibson and they in turn built their own version...the "Go Fast Jet Pack"....still another copy of a copy...of a copy. The science and technology had yet to change at all.

In 1997 Nino Amarena of San Carlos, Ca. contacted me in hopes of building his version of a bipropellant rocket belt. I helped Nino out over a period of several years and in 2007 finally free flew his "Thunderpack".

The test stand runs with the bipropellant were successful, it performed wonderfully.....but I did not want to strap it to MY BODY with temperatures exceeding several thousand degrees. We flew it solely as a monopropellant (90% Peroxide)....go figure! Nino was the first person to actually successfully build and fly a version

2-man POGO

Lunar POGO concept

Flying Chair

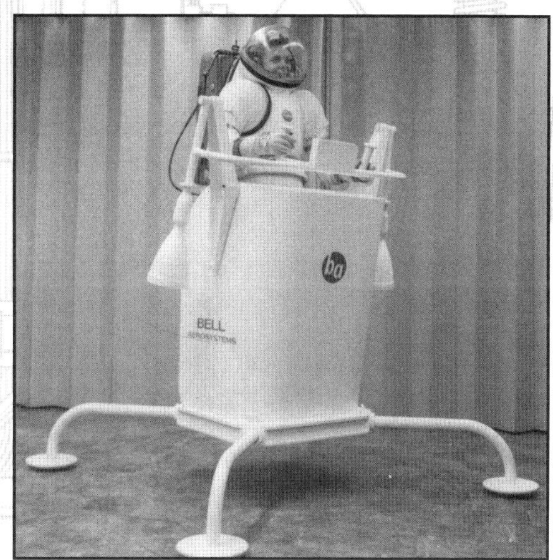

Lunar POGO prototype

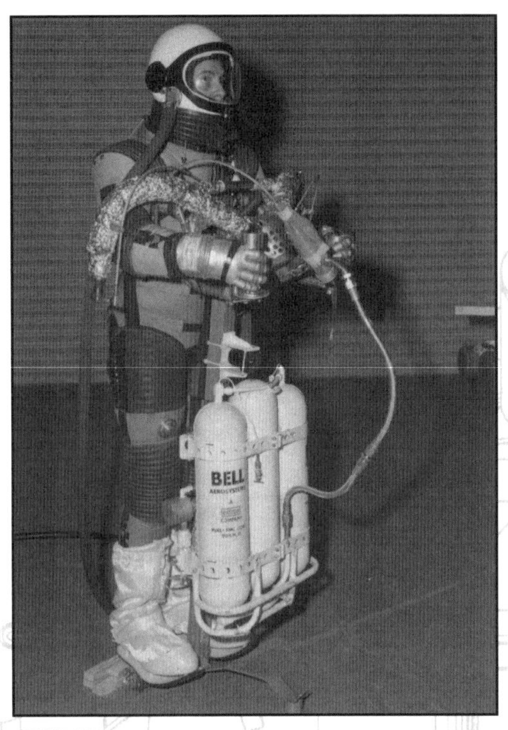

POGO with early pressure suit

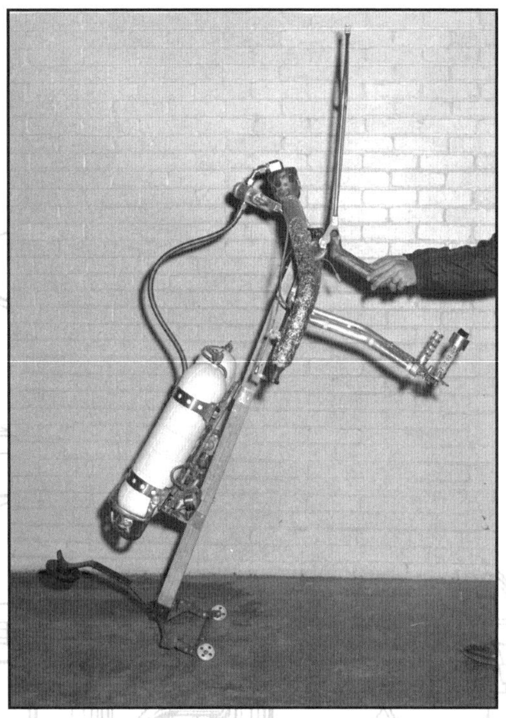

"Reverse" POGO

that was actually an advancement in the science, the rest were all Bell knock offs.

Now as I write this, Stuart Ross a daring Englishman with lots of pluck and a few quid, has just joined the "club" of "Rocketmen". Stuart has successfully built and free flown his rocket belt, again with under 30 seconds of flight time.

These individuals are the only ones I am aware of at this time who have "FLOWN FREE" (not on a tether). I may have missed someone, if I did I am sorry, but I consider "Free Flight" to be the test to determine if you have built an honest-to-goodness "Rocket Belt".

There are a number of people around the world hard at work on their own versions, God bless them, I wish them great success, but until you free fly your machine....it's still just another expensive hobby. There's Gerard Mertowlis in New Jersey who I feel will succeed by sheer will power, Juan the Mad Mexican, who if he keeps at it will have his Pogo flying soon, a young man in Arizona, and a fellow in Switzerland. Keep at it folks, but be careful.

Rocket Relationships

The simplest way to describe a rocket is to borrow a page from the Bell Aerosystems Rocket Data Handbook.

The fundamental principle upon which all jet and rocket prime movers operate is based on Newton's third law; i.e. for every action there is an equal

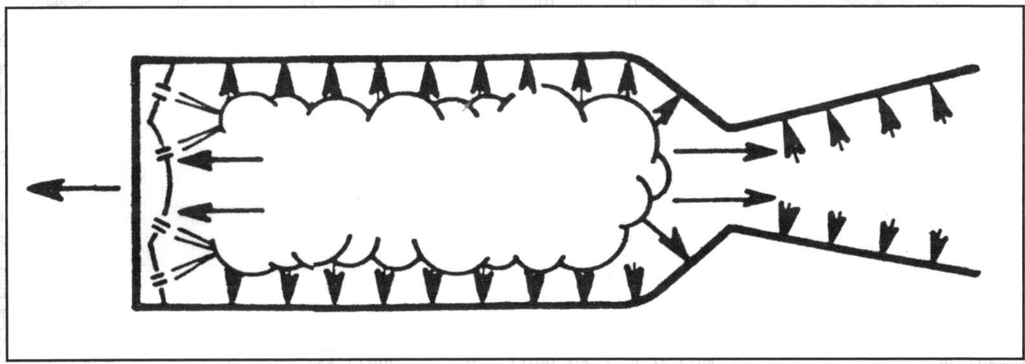

and opposite reaction. To afford a more complete understanding, consider the above illustration of a typical rocket thrust chamber.

Upon combustion of the propellants in the thrust chamber, the gases expand through the nozzle at a high velocity. The internal pressure at the nozzle end is relieved, leaving an unbalanced pressure at the other end that tends to propel the chamber or the vehicle to which it is mounted in the direction opposite to the issuing jet. Propulsion is dependent upon internal conditions alone and not the effect of the jet pushing against the surrounding air.

In contrast to other forms of jet engines, the rocket does not use the oxygen in the atmosphere for the combustion process. Instead, the oxidizer is carried aloft with the fuel. Thus, the rocket is the only means of achieving travel beyond the atmosphere of the earth.

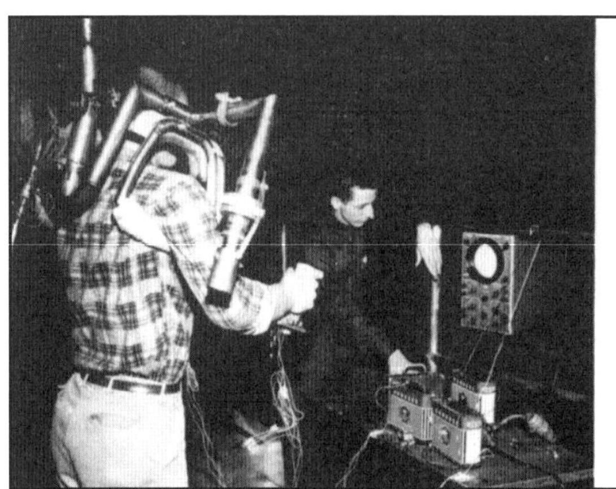
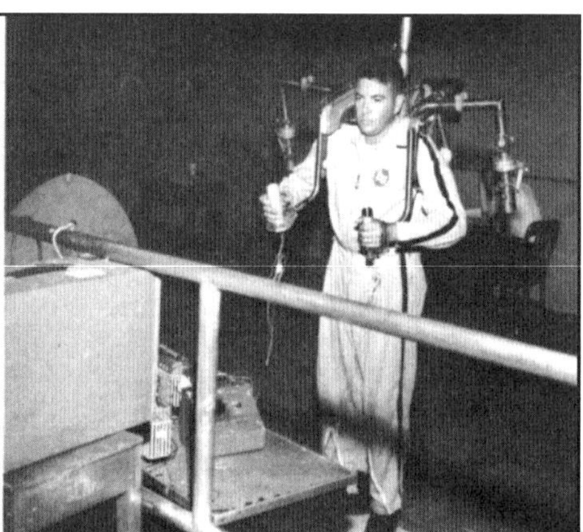

Rocket Belt Trainer, note early gimbaled nozzle.

Inventor Wendell Moore tries out a tethered prototype

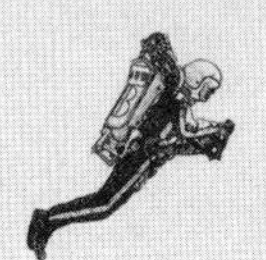

Rocket Fuel - Hydrogen Peroxide

As I stated, the Rocket Belt uses a hydrogen peroxide (H_2O_2) rocket propulsion system. This peroxide is not the 2% type found in your medicine cabinet; it is a very powerful and very dangerous 90% strength that is no longer produced commercially anywhere in the world.

Most other rockets (and jets too) use bipropellants: the fuel and a separate oxidizer. An oxidizer is a chemical that provides oxygen so that the fuel can be burned. Jets can "breathe" their oxidizer from the air they fly in, but rockets must carry the extra weight of the oxidizer with them into space.

Hydrogen peroxide is a monopropellant; it serves both as a fuel and as an oxidizer. Very simply put, H_2O_2 is water with an extra atom of oxygen, H2O + O, but oh what that extra atom can do! When it comes in contact with organic materials or precious metals (like silver), a powerful chemical reaction takes place in an unbelievably short time (two-tenths of a millisecond). There is no actual combustion, but there is the release of water, oxygen, and heat (1388° F). The heat vaporizes the water into steam that powers the Rocket Belt: A flying locomotive!

So now you know that if you combine hydrogen peroxide and a precious metal, high-temperature steam will be produced. If this is done inside a strong chamber with an opening in one side, the steam will shoot out in a jet and something is going to move. In our case, that something is the Rocket Belt and whoever is strapped to it. Every second, the Rocket Belt consumes 1½ quarts of peroxide while exhausting over 70,000 quarts of steam. That is about a milk carton full of peroxide and ten school buses full of steam, all happening faster than you can blink your eyes.

A Short History of Hydrogen Peroxide

Hydrogen peroxide was discovered in 1818 by the French chemist Louis Jacques Thenard. He made it by reacting barium peroxide with acids. Little use was made of this discovery for over 50 years. The first commercial manufacturing of hydrogen peroxide seems to have been carried out with the barium process at the plant of Schering in Berlin in 1873. An electrolytic method for making hydrogen peroxide can be traced back to the research of Faraday, who observed in 1832 that sulfate solutions could be oxidized at a platinum anode. About 1910, the Osterreichische Chemische Werke in Weissenstein set up the first commercial plant for this purpose. Sulfuric acid was oxidized in $H_2S_2O_8$ on platinum anodes, after which hydrogen peroxide was recovered, as proposed by Teichner.

In 1909, two German chemists, Pietzsch and Adolph, patented an alternate way to synthesize hydrogen peroxide via electrolytic persulfate. This process anodically oxidized an ammonium bisulfate solution, yielding ammonium persulfate. Hydrogen peroxide from the various electrolytic processes proved purer and more stable than that from the barium process. Electrolytic hydrogen peroxide was first made in the United States in 1926.

Before World War II, the Germans carried on secret projects and built facilities aimed at producing 80%-85% hydrogen peroxide for use in weapons. They also managed to produce some at 98% strength and keep it stable. These developments were not completed in time, but some (like high-speed submarines and long-range rockets) might have changed the course of that war.

The Turin Air Show - 1968

CHAPTER 6

The Engine

The Bell X-1A

Bell relied on its previous experience with rocket systems to adapt and design the SRLD system. Bell had developed similar H_2O_2 rockets for attitude control of the X-l and X-15 airplanes and the Mercury space capsule. Although these H_2O_2 rockets were too small to generate the 300 pounds of thrust needed by the Rocket Belt, design technology for these small thrusters was adequate to permit scaling to higher thrusts.

The Rocket Belt Engine

The rocket engine on the Rocket Belt is only about the size of a grapefruit, yet this mighty little chamber, machined from special metal alloys, produces well ever 1,000 hungry horsepower. As I said, it consumes more than 1½ quarts of fuel every second, not one of the EPA's runners-up for fuel economy! The engine is actually called an expansion chamber.

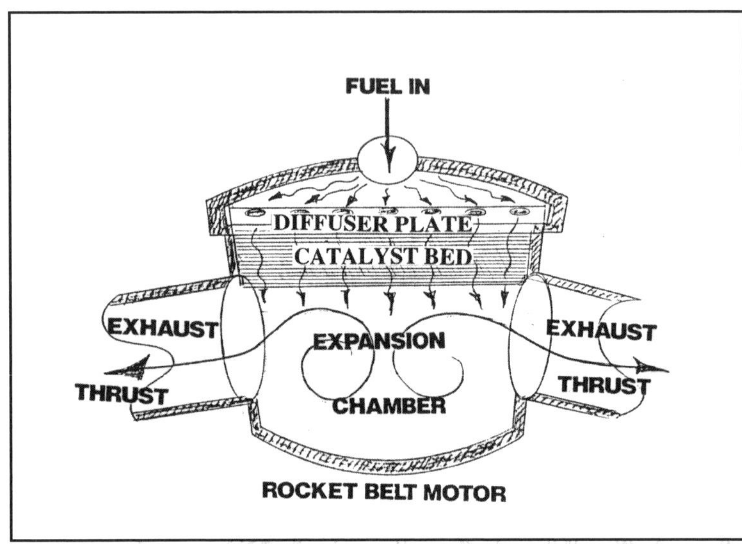

How it works.

Fuel enters the expansion chamber at about 570 psi through a tube with a restricting orifice in it. Inside this chamber is the catalyst bed: a series of fine screens layered in and compressed together very tightly. The screens are predominantly silver with small amounts of platinum and gold; a very expensive engine, indeed! Openings in adjacent screens are offset slightly at a very precise angle. Sitting directly atop the cat bed is a metal plate with a precise number and diameter of holes that evenly distribute the incoming fuel. This is known as the diffuser plate.

The "cat bed" has the same diameter as the inside of the chamber. It fits so tightly that, before it can be pressed into place in the chamber, the cat bed has to be chilled in dry ice to shrink it, while the chamber is heated with a welder's torch to expand it. Each cat bed has its own "personality" and the orifice must be changed to fit that personality and make the engine run optimally.

Two openings at the bottom of the chamber equally divide the exhaust steam into the two thrust tubes, also made of special high-strength metals. The thrust tubes are insulated on the outside to protect the pilot, and reduce heat loss from the steam. This type of flight system with split exhausts is said to be bifurcated.

We now have a rocket propelled by high pressure steam escaping from two exhaust pipes. With exhaust temperatures at 1,388 °F, we have to be VERY careful. In all the photos you have seen, or if you have been lucky enough to see one of the actual Belts, the thrust tubes appear to be covered with aluminum foil. Correct; that is foil. Underneath it is a special insulation that you could feel if you squeezed the foil. Their, primary purpose is to keep the tubes as hot as possible to keep the

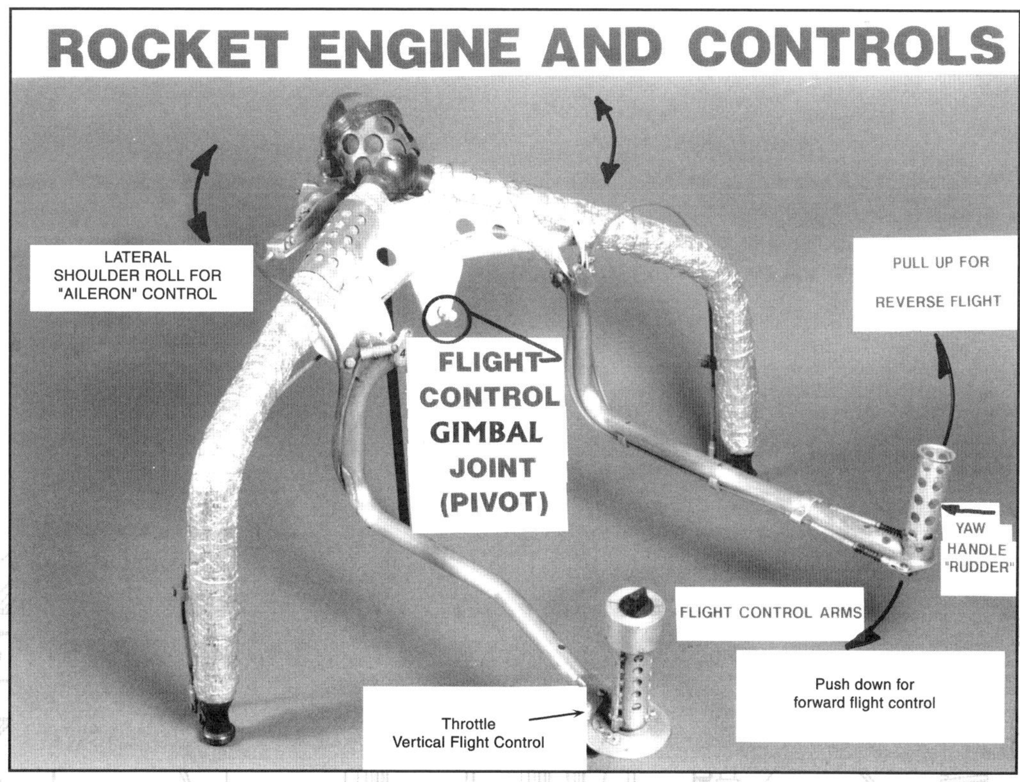

exhaust gases as hot as possible. The hotter the gases, the greater their volume; the greater their volume, the greater the pressure and consequently, the thrust. The foil also helps protect the pilot and crew from being injured while working around a hot Belt. During flight and immediately afterwards, the chamber and tubes actually glow a reddish-orange color from the heat; you can see this well during night flights.

The H_2O_2 rocket produces an extremely high noise level, 132 decibels (about the same as a jackhammer), sufficient to cause discomfort in unprotected ears. The sound is not a roar, but more like a ruptured air hose, multiplied many times.

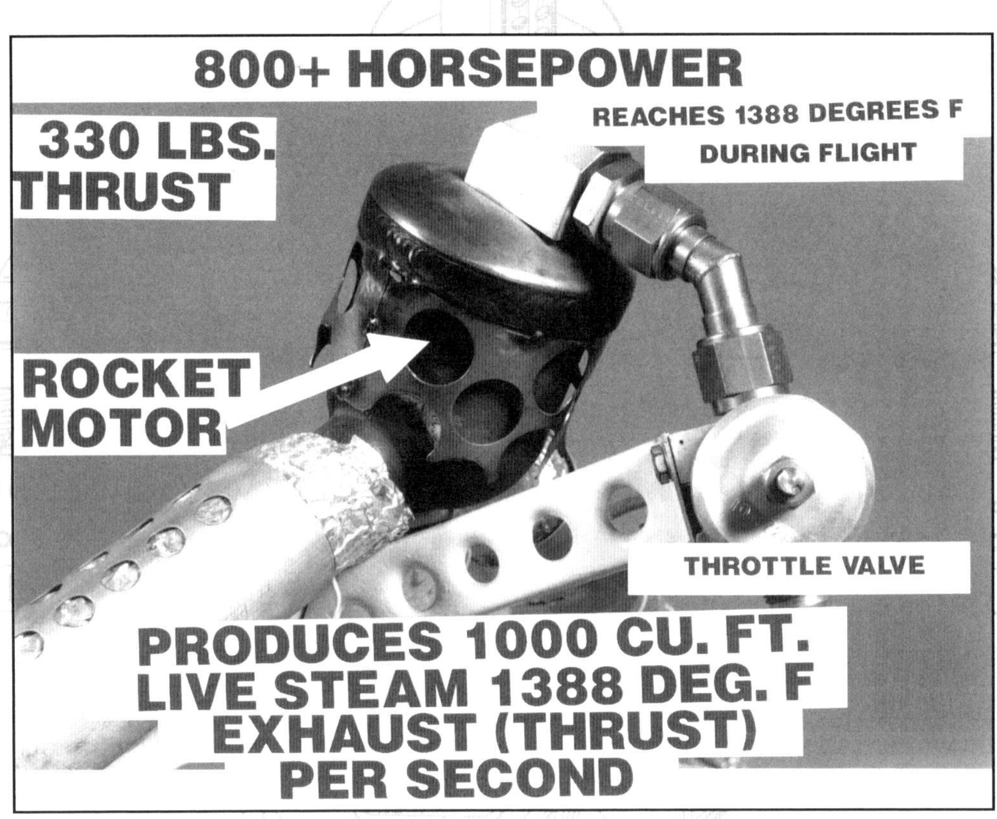
800+ HORSEPOWER
330 LBS. THRUST
REACHES 1388 DEGREES F DURING FLIGHT
ROCKET MOTOR
THROTTLE VALVE
PRODUCES 1000 CU. FT. LIVE STEAM 1388 DEG. F EXHAUST (THRUST) PER SECOND

CHAPTER 7

Evolution of the Rocket Belt's Controls

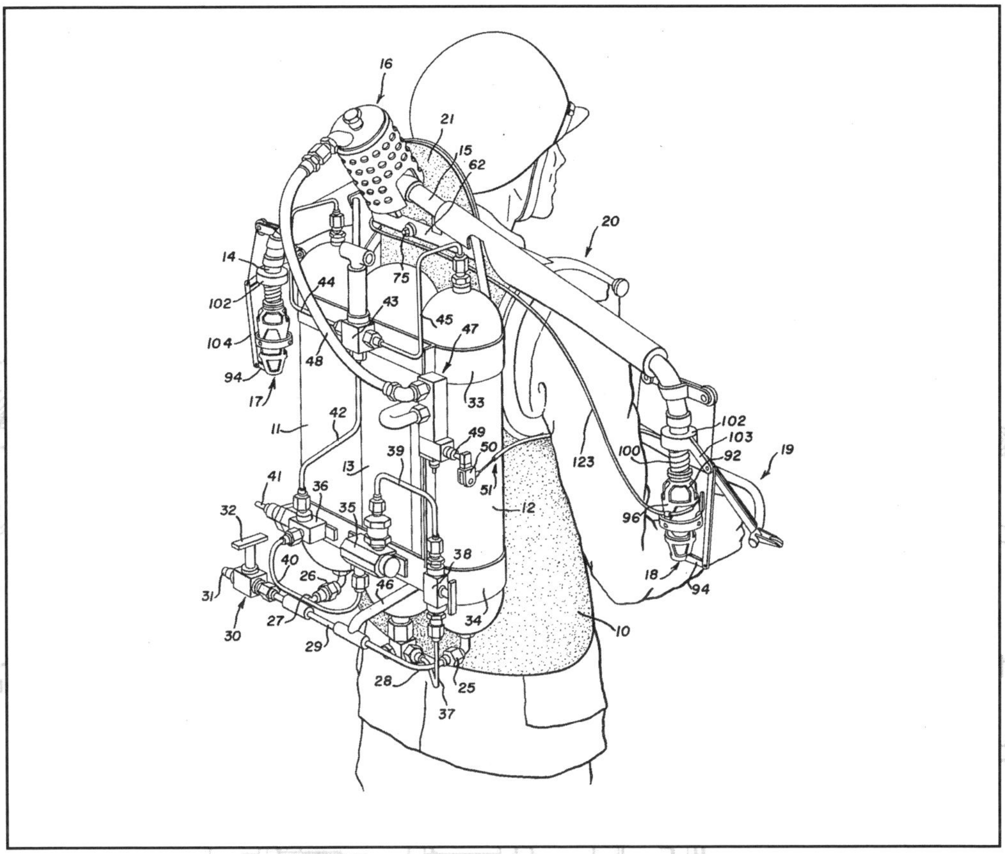

Of course, the Rocket Belt did not just leap off its designers' drawing boards in the form that we see it now. Although Moore had a definite vision for how the thing should work, nothing of this sort had ever been done before; trial-and-error was needed to produce a safely controllable vehicle. The accompanying pictures, from Moore's United States Patent Number 3,021,095 on the Rocket Belt, show his original ideas.

Altitude control would be accomplished by changing the amount of thrust being generated: increase thrust to go up, decrease thrust to go down. Therefore, some type of throttle was needed. A squeeze-type handgrip located in the flier's right hand was tried first. This initial system produced poor thrust response characteristics because it had too much travel. Also, it was discovered during training that if

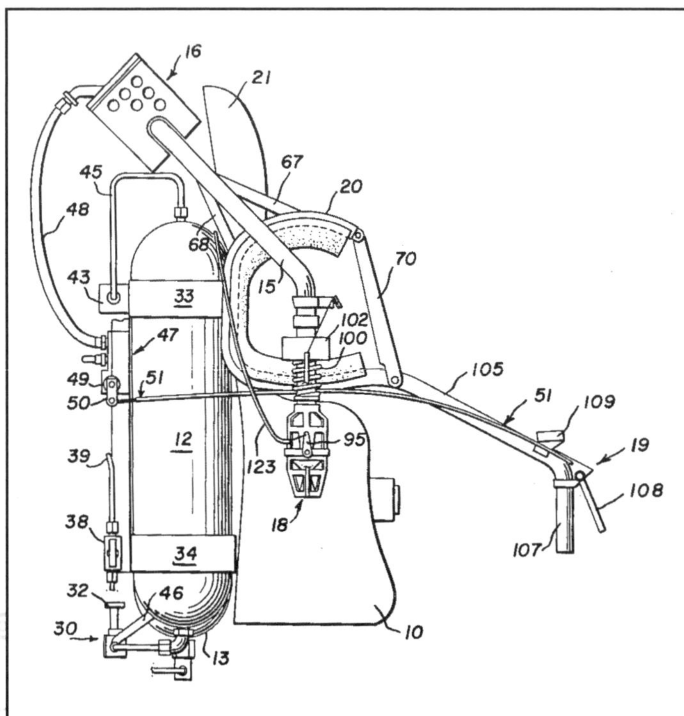

the pilot got into a "difficult" situation, his unconscious tendency was to tighten his grip. This caused him to fully open the throttle just at a time when this was the LAST thing he should do. A twist-type handgrip (like on a motorcycle) replaced the squeeze-type throttle control, and worked much better. I discuss the throttle in more detail later.

Directional control was a more difficult problem to solve. "Steering" a rocket involves changing the direction of the rocket's reaction force, known as the thrust vector. On the Rocket Belt, the thrust vector is determined by the direction of the exhaust steam with respect to the center of gravity (CG) of the man-machine combination.

In his Rocket Belt patent, Moore proposed two separate means of thrust vectoring for controlling horizontal translational flight, with stability provided by the pilot's instinctive body motions (i.e. kinesthetic control). The design of the flight control system included:

A three-axis control stick whose handle could be simultaneously rotated and tilted longitudinally and laterally by the pilot's left hand. Each rocket nozzle was mounted on a special ball-and-socket joint called a gimbal. Movements of the control stick caused the nozzles to be tilted, thereby pointing each nozzle's thrust vector in the desired direction. The nozzles were restricted from pointing inward, to keep the jet blasts away from the flier.

The capability of longitudinally and laterally tilting the upper assembly (gas generator, lateral gas supply tubes and rocket nozzles) of the propulsion system was through movements of the pilot's arms. A longitudinally-oriented pivot on the harness permitted lateral

tilting; longitudinal tilting was by fore-aft flexing of the upper extension of the harness.

Difficulties with the nozzle gimbals arose during tethered testing on the initial Rocket Belt system. With the upper assembly pivoting locked (so control was solely through nozzle tilting), the three-axis stick system was overly sensitive to the pilot's control inputs. Further, the nozzle gimbal system did not maintain good neutral alignment.

The stick system was replaced by the control arm approach. Two control arms are attached to the thrust tubes, enabling the pilot to change the angle of the tubes. These control arms pass under the pilot's armpits and extend straight out in front of him so that his arms are 90° to his body. The throttle handle is at the end of the right control arm; the handle on the left control arm controls yaw.

A less sensitive, universally-tiltable device called a "jetavator" located at the nozzle exit; replaced the gimbaled nozzles. What are jetavators? Jetavators are small metal rings, slightly conical in shape, that are mounted so that they pivot slightly and dip into the exhaust blast simultaneously and at opposing angles, causing rotation of the Rocket Belt (and pilot) about the vertical axis. The jetavators pivot on a set of pins and have a small control arm, that faces out from the aft side of the rocket nozzle.

Finally, to lessen control sensitivity, the thrust tube gimbal was relocated so as to create a 30-pound upward force on the control arms. This created a sense of pressure the pilot could constantly "lean on" to help make smoother arm movements; a sort of artificial gravity for weightless pilots.

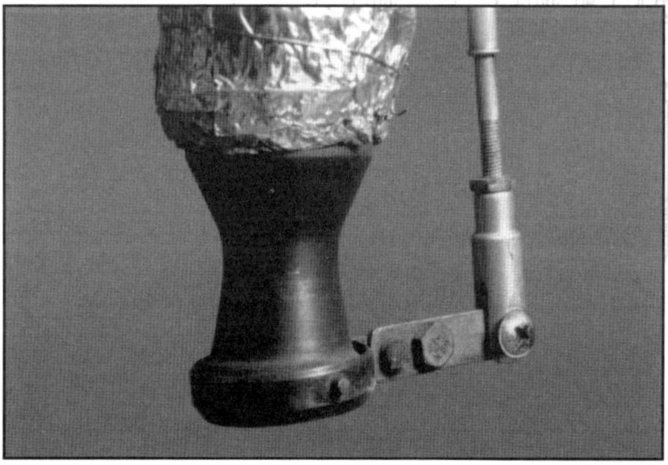

With these changes, the feasibility demonstrator was able to make free flights and satisfy flight operations requirements of the contract.

Jetavator

Bill Suitor - 1982 World's Fair

CHAPTER 8

The Corset and Straps

Let us see how our little engine is put to work to overcome gravity. In order to fly, the-rocket has to be attached to the pilot's body, keeping in mind that it is over 1000°F. You cannot very well just sew a rocket to someone's clothing.

Therefore, it is mounted in a metal framework that is strapped to the pilot's body. This framework is called the backbone. The exhaust pipes have to be long enough to carry the steam away from the pilot's body and must be positioned so that the reaction forces are upward through the air.

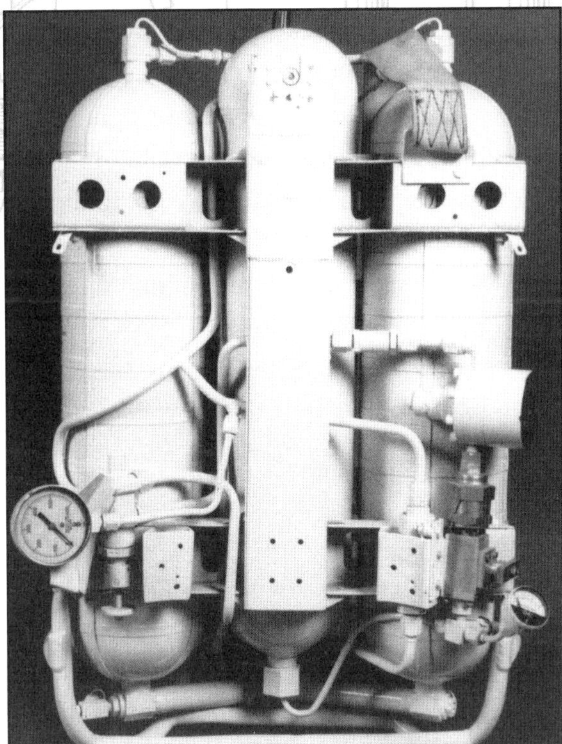

The metal frame can be seen in this image.

Two important factors must be kept in mind now that this is attached to you. The first is that it weighs 110 pounds when fully fueled. If it merely "hung" on your back with simple shoulder straps, in a very short time you'd be very uncomfortable, if not in downright pain. Then, if you had to carry this load even a short distance, you would not be too sure of your footing. When walking around with a half million dollar, fully fueled and pressurized "bomb" hanging on your back, you must be very certain of your footing. Your "landing gear!"

The second and most important feature is that the system the pilot uses to carry the Belt on the ground will carry him in flight. If an arrangement like a parachute harness were used, you wouldn't fall out, but ... the tremendous power of the rocket would cause it to creep up your back until the control arms would be jammed up

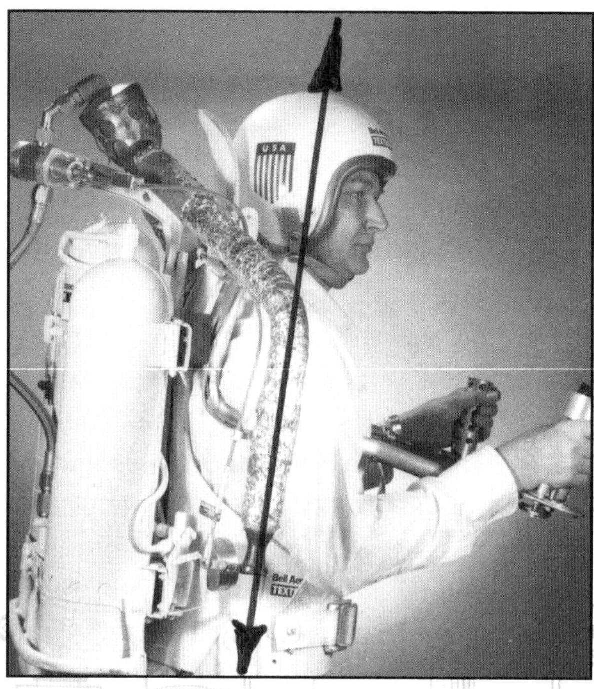

To keep from tumbling the pilot must put the thrust through the centre of gravity (CG) of the man/machine combination.

into your armpits. This would create some problems:

o During flight, most of his weight would hang from the Rocket Belt control arms; this would make smooth control almost impossible.

o The control arms would greatly reduce the blood flow to your arms and hands, causing loss of the light touch needed to control the Rocket Belt.

o As the machine crept upward, the straps between your legs would become very tight and (among other painful things) the blood flow to your feet would be cut off; hence, the loss of feeling in the feet. The loss of control of the "landing gear."

The inventors at Bell put a tremendous amount of human-factors engineering into these problems. Many long hours of brainstorming and testing were needed to create a corset suitable for the multi-role job of "pilot carrying Belt" and "Belt carrying pilot." What they came up with is a unique device known as the Bell Hip Pack, or "Hipack". This device transformed man into a beast of burden by transferring loads from his back to his (stronger) leg muscles. Rather than get into a long explanation of the Hipack, allow me to show some of the various jobs that it could accomplish. I'm very sorry to say, however, like so many of Bell's ideas, the Hipack was another first that never got off the ground.

It is a fiberglass corset lined with a layer of dense foam padding that wraps around the midriff, closing with an overlapping belly-band single harness strap and buckle. It conforms to the contours of the back and extends up to the top of the head. The Rocket Belt

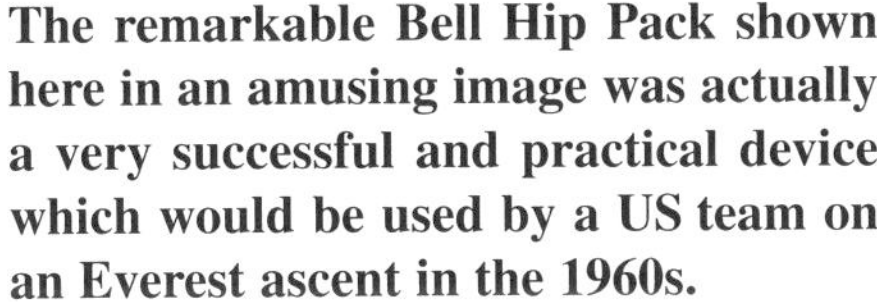

The remarkable Bell Hip Pack shown here in an amusing image was actually a very successful and practical device which would be used by a US team on an Everest ascent in the 1960s.

backbone is securely attached to this corset. A chest strap is attached to the backbone at the upper cross-member; this strap holds the corset (and therefore the Belt) tightly on the pilot's back. Two leg straps, attached to the lower portion of the corset, fasten much the same as a parachute harness. These leg straps, when properly adjusted, keep the corset from creeping upward on the pilot's body and do not support much body weight in flight.

Therefore, the rigid fiberglass corset does the following jobs admirably:

* Securely attaches the Rocket Belt flight apparatus to the pilot's body to ensure a proper fit, enabling the man-machine to become the "Man-Rocket."

* Positions the majority of the payload on the pilot's hips, thus putting the load on the major leg muscles and keeping it centered for good balance.

* Provides a common surface for the secure attachment of all harness straps.

* Allows the pilot to carry the Rocket Belt comfortably for long periods.

* Affords a high degree of protection to the pilot's body and spine.

Once the basic idea was selected, the next step was making hardware. An average-sized person was selected to lie in a tub of plaster to create a mold for the corset. To the right is a picture of that procedure.

Now, something is needed to hold the pilot securely in the corset. At its inception, the corset had the front open, so the Kelly Belly Plate was devised. It was named after Dr. Francis Kelly; the Bell physician. Although the belly plate served its purpose, it proved, a bit cumbersome to hook into.

With the evolution of the corset came the overlapping bellyband. The subsequent wrap-around band proved more comfortable and gave the wearer a feeling of being surrounded and more secure. As you can see from the photos of the Rocket Belt corset, the bellyband is wide and overlaps in the front, is lined with firm foam, and is secured with a single strap and buckle. Its large area distributes the load on the pilot. I would best describe it as a giant pair of hands lifting you up about the waist.

A chest strap holds the upper part of the Belt against your back. It attaches to the backbone cross-arm on the left side and crosses the chest diagonally to a buckle attached to the corset on the right side, sharing a mounting bolt with the belly strap.

Peter Kedzierski wearing the Kelly Belly Plate (1962)

On the original Bell "A" Belt, the cross-chest strap was not used. In its place were two latching bars attached to the control arms that folded open and down. They were padded and held the pilot into the Belt. Maybe a better way to describe it was that they held the Belt against the pilot's back. All the padding made free and easy movement of the control arms difficult; there was never a good way to lock a man in. Improvements were always welcomed and with the Bell "B" Belt came a set of rings that buckled around the pilot's arms. These were an improvement but again the heavy padding under the pilot's arms interfered with free control-arm movements. The final arrangement on the late model "B" Belts was a single cross-chest strap.

This finally brings us to the leg or crotch straps. Their main function is to hold the Belt down around the pilot's waist. Without them, the Belt would creep upward as thrust is applied, putting the pilot's weight on the control arms. On the original Bell "A" Belt, a pair of straps was used with a rubber sheet ("diaper") between them; this was eventually eliminated.

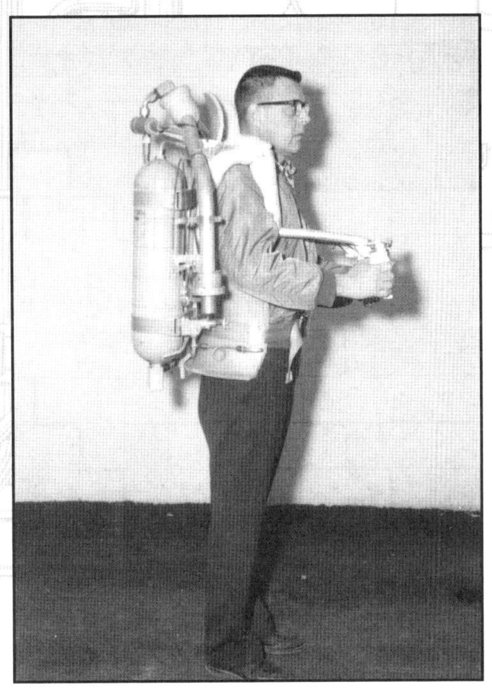

Wendell in the original mock-up. Note early control handles and gimbaled nozzles.

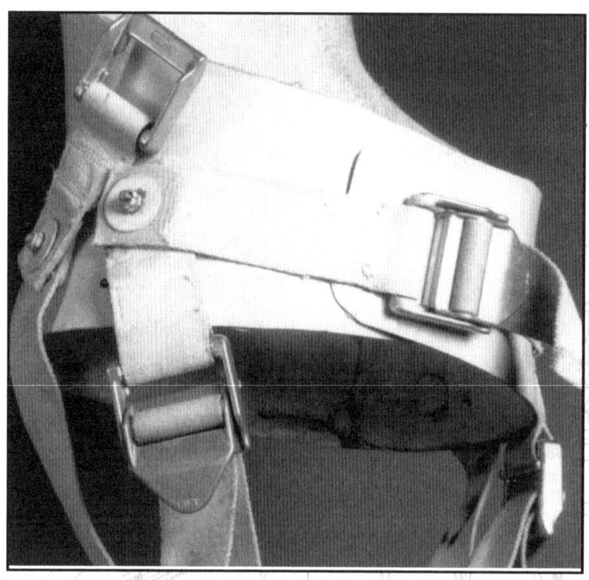

The leg strap assembly

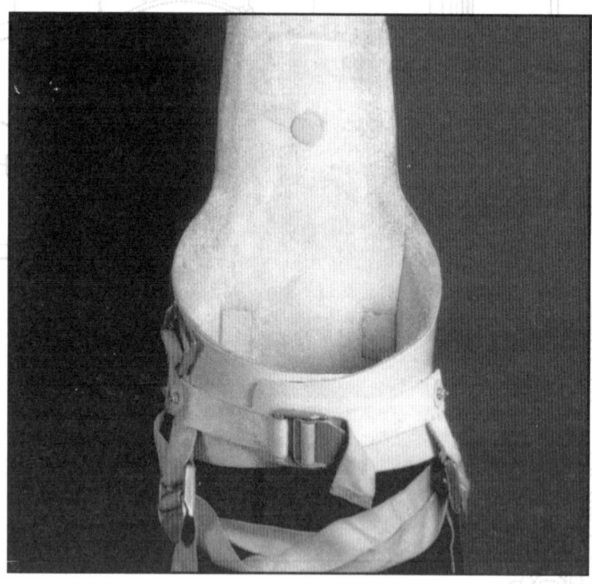

Front view of Hip Pack

Of course I need not advise that the pilot's privates should not be left under a strap, to be pinched against his body as he takes off; very distracting, let alone painful!! When flying a Rocket Belt, you really don't want any distractions, and certainly not any of that nature!

Strapping into the Belt

There is a three-legged stand that supports the Belt during fueling, servicing, and while the pilot is being strapped in. It is a horseshoe-shaped platform that surrounds the pilot's hips with the opening facing forward so the pilot can back into the corset. Two stabilizer arms, mounted on each side at the front of the horseshoe, connect to the Belt at two points, one on each side of the backbone's upper cross brace by means of two quick-disconnect pins. The stabilizer arms keep the Belt from tipping over and falling off the stand. The stand's legs are adjustable to accommodate different pilot heights.

As the pilot steps into the Belt on its stand, he bends his knees slightly and, as the bellyband is closed, he has to sense the most comfortable position about his waist. He checks that he has the proper clearance between his armpits and the con-

The three-legged support stand

trol arms. When the Belt is in the proper position, the bellyband is tightened and the chest strap adjusted so the Belt is snug against his back. Now he can disconnect from the stabilizer arms and stand upright to adjust the leg straps properly. When you are properly strapped into the Belt, the control arms should be slightly beneath your armpits, allowing the control arms to tilt smoothly up and down. Without the leg straps, the distance between your armpits and the control arms would not remain constant as thrust lifts the Belt.

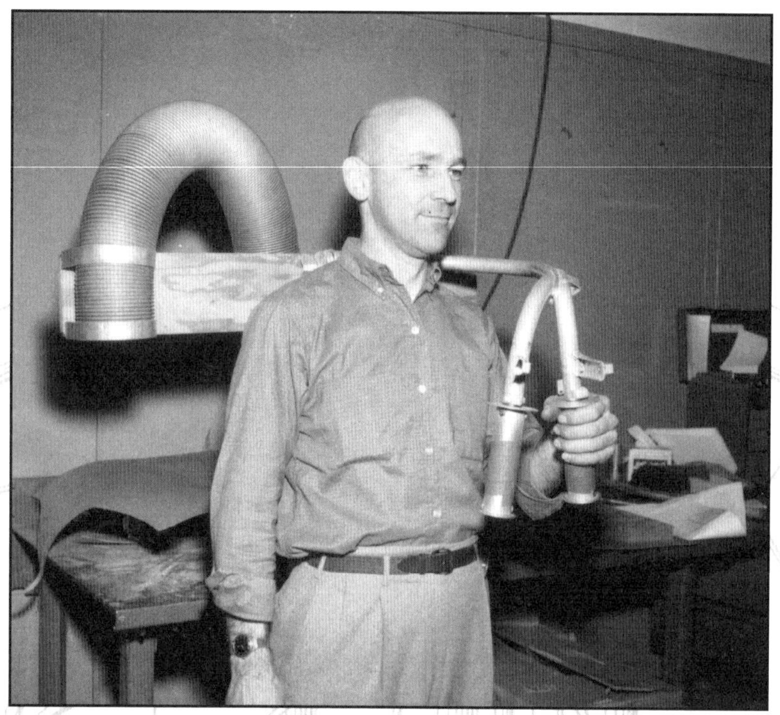

Single arm prototype

CHAPTER 9

Fuel System

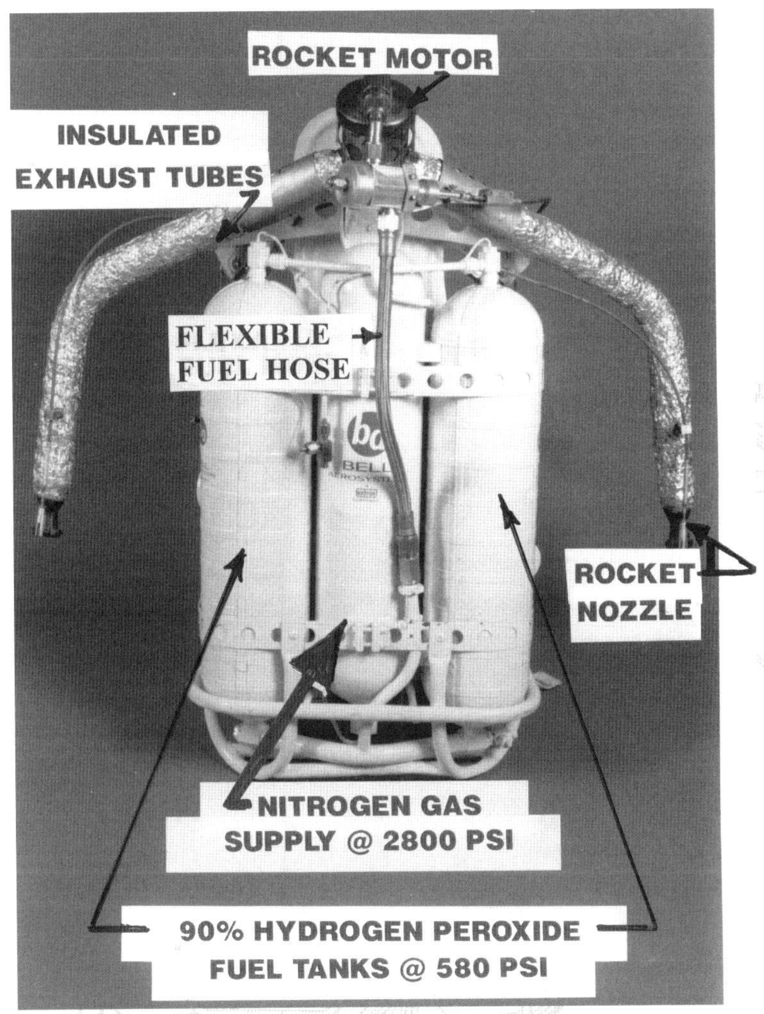

The Bell Rocket Belts are limited to 21 seconds run time, consuming approximately 50 pounds (6 gallons) of hydrogen peroxide fuel during that time. The fuel is stored in 2 tanks mounted to the backbone of the Rocket Belt. These tanks are made from stainless steel for strength, and must first be "pickled" or passivated in a process that makes them not react with the peroxide fuel.

Between the fuel tanks is a high-pressure gas cylinder containing nitrogen. It is a wire-wound, high-pressure, shatterproof "bottle" that resembles a

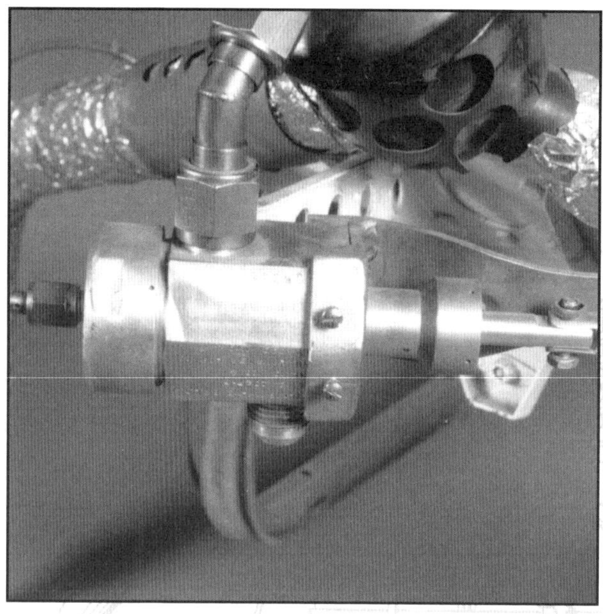

Throttle valve

scuba diver's breathing apparatus.

To get the H_2O_2 fuel from the two outboard tanks, the N_2 is routed through tubing to a gas regulator that reduces the pressure from the higher storage pressure to the lower working fuel pressure. The fuel, now at working pressure, is forced from the tanks through a series of valves and high pressure tubing into a common manifold, then up through flex tubing to the throttle valve.

The throttle valve meters the fuel that is forced into the catalyst bed and thrust chamber. It is rigidly attached to the thrust tubes by a mounting bracket; the throttle control cable runs from the throttle valve to the throttle handle that attaches to the right control arm. The throttle valve moves with the thrust tubes, hence the need for the flexible fuel line. The throttle valve, manufactured specifically for the Rocket Belt by the American Waterlift Company, is truly a unique device and is the most important component of the entire system.

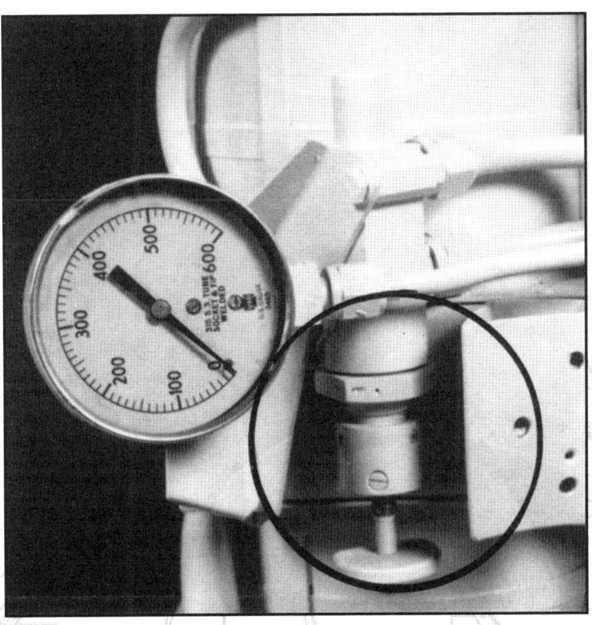

Nitrogen shut-off valve

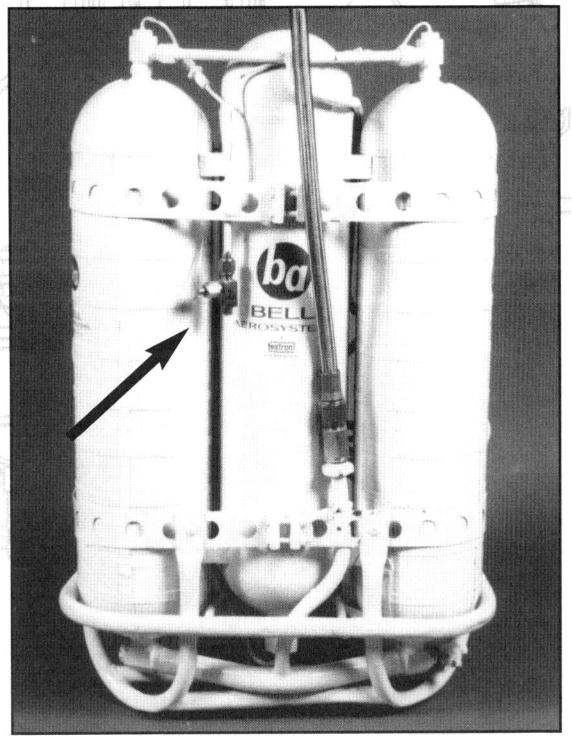

Fuel vent valve

Check valves, flow valves and gauges complete the fuel system; it is important to be familiar with all the valves and gauges and their locations.

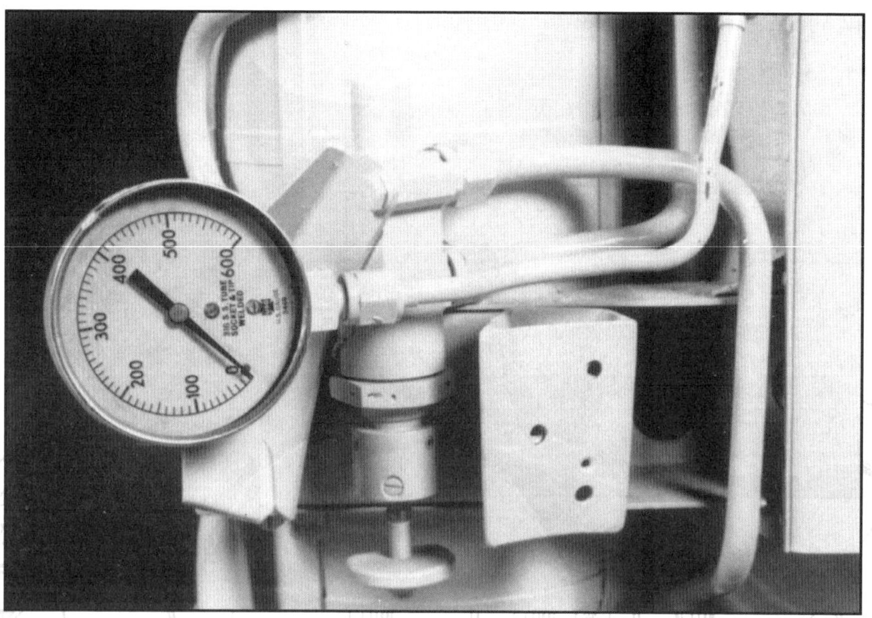

Fuel Pressurization valve

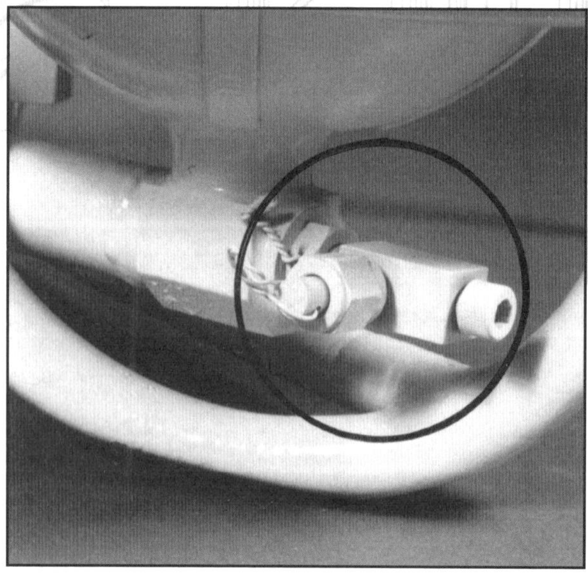

Fuel fill valve

The Throttle

Now let us move on to the throttle and how it functions. The throttle handle is located at the end of the right control arm. It is a motorcycle-type twist handle that rotates counter-clockwise for power increase. A removable safety locking pin goes through the throttle handle to keep it from being twisted accidentally. Inside the throttle handle is a battery for the fuel warning buzzer; atop the throttle handle is a twist type timer that tracks flight time and activates the warning buzzer. I will cover these in detail in another section.

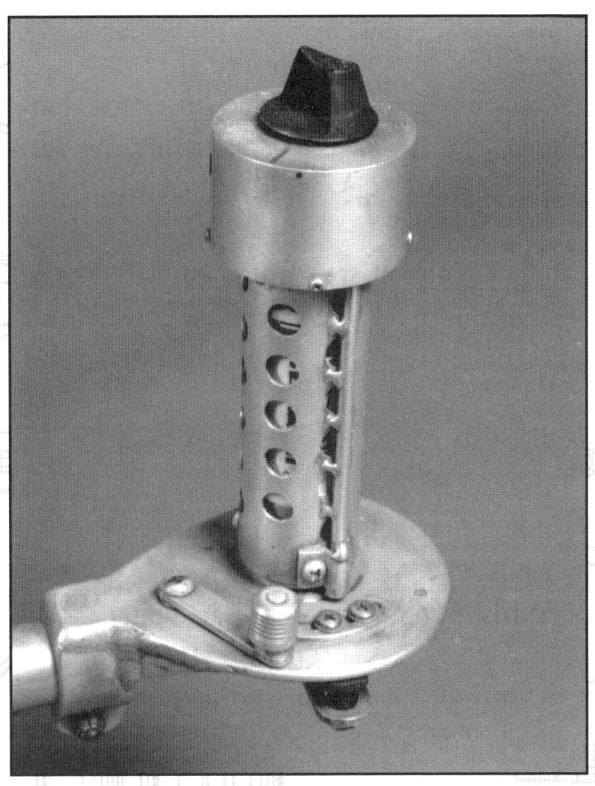

The Throttle

The throttle has a very interesting feature: its MINIMUM setting is 60% of full power; strange, but let me explain why. Keep in mind that a fully fueled Belt weighs in around 110 pounds and the typical pilot at around 175 pounds, for a total "payload" of 285 pounds. Any amount of thrust less than 285 pounds will not lift the payload off the ground. A Rocket Belt pressurized to a working pressure of 570 psi will produce approximately 330 pounds of thrust, giving the pilot about 45 (330-285) pounds of thrust to work with at takeoff.

The human wrist is limited to about 90° of rotation while in the vertical position. If we were to have a throttle range from 0% to 100% spread out along that 90° of rotation, the actual "usable" portion would be only in the last 40% of travel. The majority of twist would be spent before the thrust reached 285 pounds; this would make the throttle so sensitive (especially later in the flight when much of the fuel weight is gone) that the Belt would be almost impossible to control smoothly. So the usable 40% of the throttle is all that is needed to control thrust over the required range and, in turn, lessen the degree of sensitivity.

Let's look at some numbers to see how the 60% throttle works and why it is so necessary:

* Pilot and Belt with full fuel load = 285 pounds
* Maximum thrust available = 330 pounds
* Thrust at minimum throttle setting = 198 pounds (60% of 330)
* Thrust which can be throttled = 132 pounds (330 - 198)

So, with 90° of throttle twist to control 132 pounds of thrust, each 1° of twist produces about 1½ (132 ÷ 90) pounds of thrust.

If the full 100% range of thrust was available in that 90° of twist, we'd be getting more than 3½ (330 ÷ 90) pounds of thrust for each 1° of twist. Remember, you can only use the range above 285 pounds of thrust, so to get to this point you'd use most of the available wrist motion before ever getting off the ground. Once there, it would be almost impossible to control smoothly. With that sensitivity, landings would tend to bounce up and down and could be very dangerous with a low fuel load. Now, I think you can see that if you spread the working range of thrust over 90° of twist it gives a much greater range of wrist motion and a much smoother power curve.

The most crucial throttle setting for a pilot to remember is the location of "off" in relation to the position of his wrist. He must NEVER inadvertently go to the "off" position while in flight. Remember, the anvil.

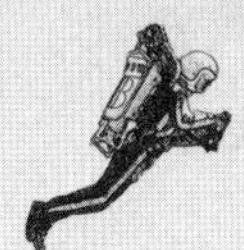

The Timer

A BELL Rocket Belt can carry only 21 seconds of fuel. So how do you know when you start running low? You certainly, do not want to be 90 feet above the ground, going 60 mph, only to discover there's only 3 seconds of fuel remaining. Remember the impact limit of sneakers!

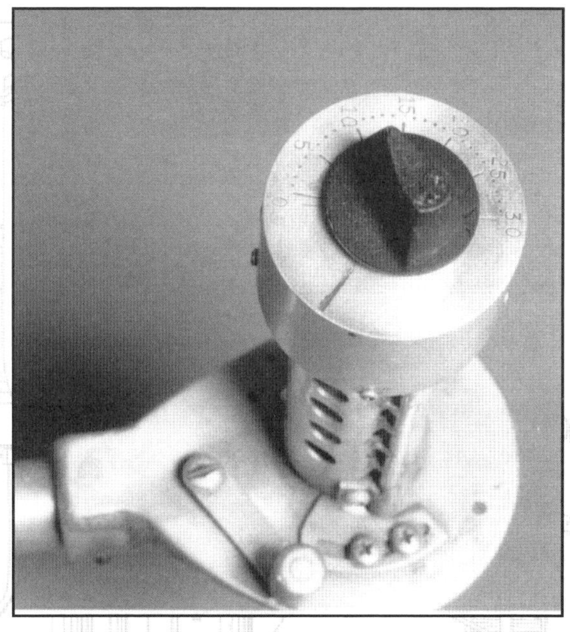

The Timer

With that in mind, a redundant low fuel warning system was developed for the Rocket Belt. A combined visual (flashing red light mounted to the front and side of the helmet) and audio (sound generated inside the helmet) system was tested first. It was not foolproof because of the rocket's high noise level and the possibility of mud being splashed on the goggles, blocking vision. The system finally selected was a battery-powered vibrator installed at the back of the helmet, driven by an automatic timer. Given the noise level and heavy "brain load" while flying, it proved to be the most noticeable system.

The timer on a Bell Belt is on top of the throttle handle. It resembles a kitchen timer, and begins to run when the throttle is opened. It is primarily used to drive the buzzer system, or vibrator, located in the back of the helmet. This device is attached to an elastic strap and fits tightly against the base of the skull.

Before takeoff, the pilot twists the timer past the 30-second mark and then returns it to the 21-second mark; this properly sets the timer. Then, when the throttle is first twisted, a small probe on the throttle handle starts the timer

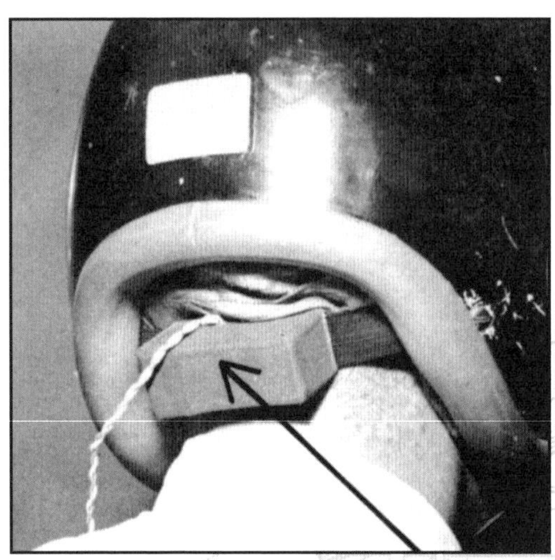

Buzzer assembly

running. After 10 seconds, the timer activates the buzzer at 1-second intervals for the next 5 seconds. Then, at 15 seconds, the buzzer begins vibrating steadily and the pilot had better be almost home.

The advice given to all Bell pilots was that after the timer passes 21 seconds, there will be silence. If you were not paying attention, it could be a silence that lasted forever! The rule-of-thumb was that when you heard or felt the first buzz, you would know if you needed to hurry or not; when you passed from intermittent to solid buzz, you had better be almost on the ground. Pilots were also instructed that if they sensed that they should be hearing a buzz while flying along, by all means look at the timer; should the buzzer fail, at least there was a visual warning device. Also, if in doubt, LAND IMMEDIATELY. Any landing you walk away from is a good landing!

Never forget that the Rocket Belt's engine burns 2½ pounds of fuel per second. This will get you into a lot of trouble in a very big hurry. Never let the "1,000 hungry horses" of that engine run away on you; never loosen a tight hold on the reins. Know where you are in relation to the remaining fuel supply at all times. It is the primary flight "barometer" by which to judge and execute all maneuvers.

Without fuel, the pilot becomes a falling object! Uncontrolled! 300 pounds of space junk falling from the sky - nothing more! Kicking and screaming won't help. Your PRIMARY RESPONSIBILITY is to those on the ground; to ensure their safety, never run out of fuel. Of course, under the heading of self-preservation, the pilot's well-being directly corresponds to that of the machine.

"Remaining fuel supply" is, to Rocket Belt flights, as "runway ahead" is to airplane landings. While coming in for a landing in an airplane, all the runway in the world does you no good when, its behind you. So it is with the Rocket Belt. If a pilot is airborne and most of the peroxide is gone, all the air in the world is useless if it is beneath him. Know as you go!

CHAPTER 12

Basic Flight Maneuvers

You, the trainee pilot, strap into the Rocket Belt. It has to fit so that you are not hanging on the control arms as it begins to lift. If this happens, movement of the controls would be difficult and the flight erratic, as well as uncomfortable. Remember, this is to be enjoyed, not endured. So, with the machine positioned comfortably, you are ready to learn the basic flight maneuvers.

It must be stressed that it is seldom (if ever) necessary to crank hard on the controls of the Rocket Belt. Any hard movement can cause violent control reactions due to the tremendous amount of power at the pilot's command. Remember, with "action-reaction," you are driving a 300-pound machine with 1,000 horsepower. A typical car weighs 3,000 pounds and has a 150-horsepower engine! It is very easy to over-control and over-correct, so the Belts have to be flown with a light touch.

Peter Kedzierski hovering

Taking Off to a Hover

A flight instructor stands to your side and tells you when the thrust tubes point straight up and down. Memorize where this "takeoff" position is.

Now, in takeoff position, SLOWLY open the throttle by twisting it counterclockwise with your right hand, adding only enough power to get about 18 inches off the ground. This keeps you out of "ground wash", the rocket blast bouncing off the ground and hitting you like a blast of gusty wind. When airborne, you are essentially weightless, and very small forces can and will effect you

Landing from a Hover

This is simply the reverse of taking off. Slowly twist the throttle clockwise to reduce thrust. It must be stressed that, during landing, the instant you feel your toes touch the ground that the power be "chopped." Remember, you are most likely 40 pounds lighter than when you took off, have 70 pounds on your back, and are putting out at least 198 pounds of thrust. If you don't cut power instantly upon impact, the thrust will bounce you back into the air (action-reaction) and then the weight on your back will start to pull you over. Your reflex reaction will THEN be to cut power ... too late ... WHAM! ... right on your butt! It hurts, you're very likely to damage the machine, and the high-pressure tanks might rupture. It's also very embarrassing.

Forward Motion

So you've taken off to a hover, and are (in theory) hovering motionless. The thrust vector is directly vertical and you are, in a sense, weightless. How do you move? You will not begin to move unless something moves you, and, being weightless, that is not going to require very much effort. You must change the thrust vector to move; to do this, you use the control arms.

To move forward, GENTLY press DOWN on the arms; this, tilts the rocket tubes slightly to the rear, changing part of the thrust vector from vertical to slightly aft. Remember, action-reaction: with the aft thrust vector (action), you will move forward (reaction).

To stop, simply relax this down pressure on the arms. They raise back up from the built-in control force and give you reverse "braking" action. As you stop, reduce power and land; simple, HA!

If you held in the downward pressure, you would continue to accelerate forward but would notice a loss of altitude. This is because some of the vertical thrust has been diverted for forward power. So keep in mind that whenever you press down on the control arms to move forward, you must also add power (via the throttle) to make up for the vertical loss. Helicopter pilots have to learn to do the same thing.

During forward flight a pilot soon comes to realize that it is not necessary to hold the downward pressure. As you might know, things set in motion tend to stay in motion. Once airborne, the only thing that slows you down is air resistance against your body. During practice, you'll learn that, for low-speed flight, it is only necessary to

"pulse" the control arms for short periods, and then coast between pulses.

Recall that those 1,000+ hungry horses consume, 2½ pounds of fuel per second. Another small but interesting effect you will notice is that, if you hover without changing the throttle setting, you will rise due to the fuel loss. All these little things come into play with one another. The loss of vertical thrust due to forward vectoring is generally offset by the decrease in overall weight due to fuel burn; mind you, this is noticeable only at very low speeds.

As an aside, if a pilot graduates from the fuel-hungry Rocket Belt to a fuel-efficient air-breathing Jet Belt, this in-flight weight loss due to fuel consumption would be negligible.

Yaw Control

As I mentioned, in the right hand is the throttle, which controls up and down movement; in the left hand is the yaw control. Yaw is a rotation about the vertical axis. Simply put, it is the ability to spin left or right: the "rudder." The yaw handle is spring-loaded so that it always returns to the neutral (straight-ahead) direction. If twisted to the right (clockwise), the Belt will turn right; twisted to the left (counterclockwise), the Belt turns left.

When the yaw handle is twisted to the left, the cable on the right side of the control handle is pulled and the

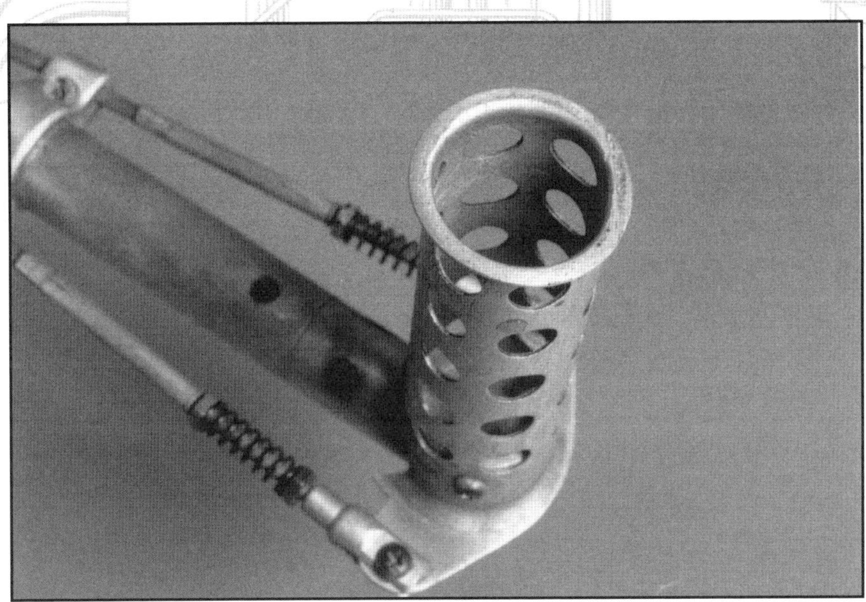

Yaw control

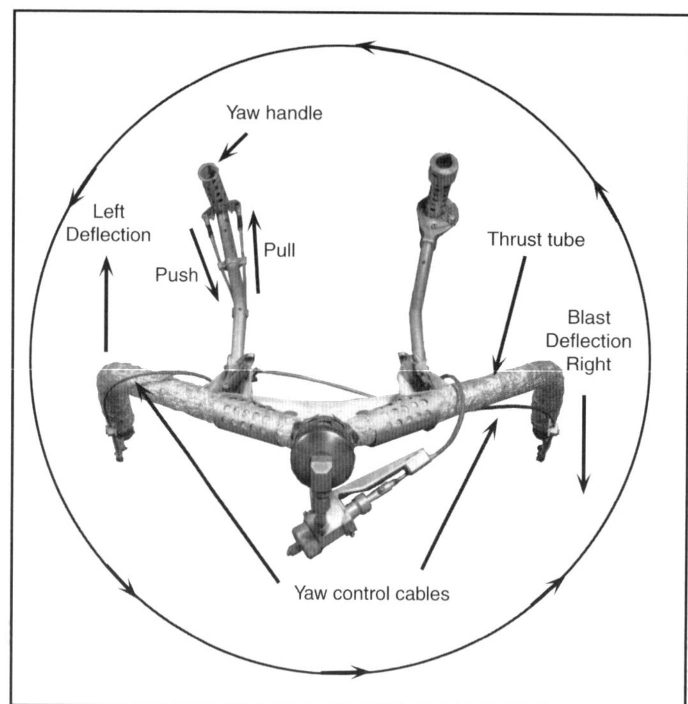

return to neutral (due to the return springs) so the need to steer constantly is eliminated.

Turns

So now I have shown how the rudder causes the pilot to face left or right. Keep in mind that the Belt is not going to go in that direction until the pilot puts in a control action to move in that direction. Let me explain.

cable on the left is pushed so as to cause the right jetavator to face (dip) slightly aft and the left jetavator to face (dip) slightly forward. Now, as the blast on opposite sides is deflected in opposite directions, the Belt and pilot are consequently spun in the direction the yaw handle was twisted. Only a tiny amount of vertical thrust is being diverted to cause this rotation, so no additional throttle need be added. When the pressure on the handle is released, the handle and jetavators

Try to picture yourself flying due east at 40 mph. Now you decide to turn due south, which requires a 90° right turn. Twist the yaw handle and pulse in a right-face command. What happens? You're now FACING south, but still flying east.

Remember, you are "weightless" and must overcome all previous directional inputs each time you wish to change direction. This does not mean

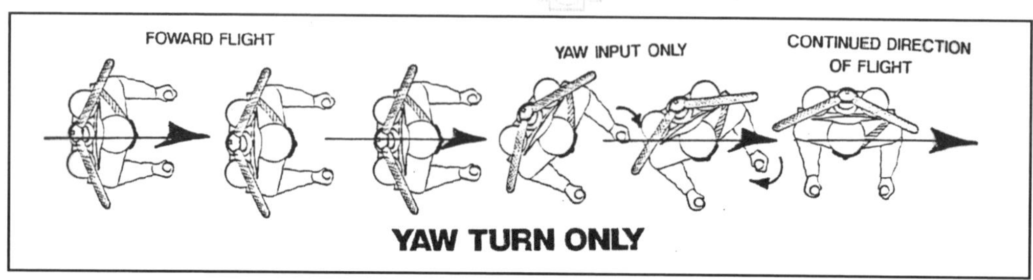

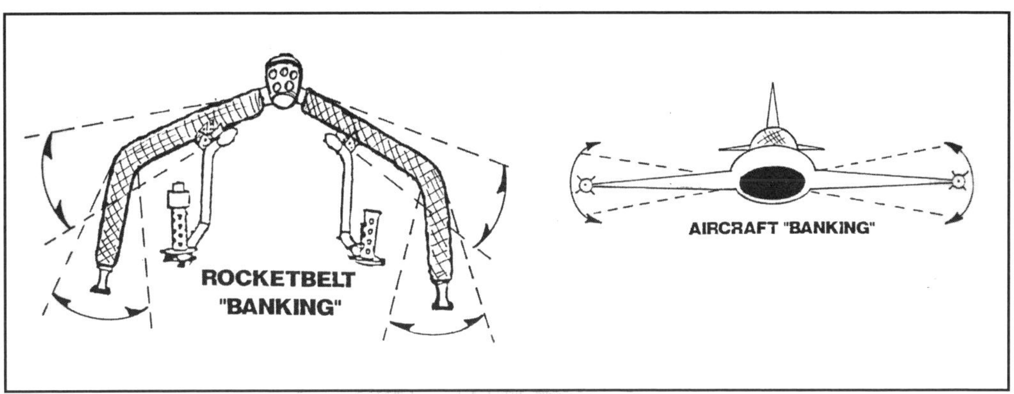

that each time a pilot wishes to head south while flying east, that he must stop, turn south, and then start forward again. What it means is that he must overcome his easterly motion as he simultaneously faces south and puts in a forward force to head south. That is called a coordinated turn.

Doing this in a fixed-wing aircraft is known as making a banked turn. That is, the pilot of the airplane puts in a lateral control that raises the craft's outside wing and lowers its inside wing, thus banking. Then the pilot uses the rudder to turn the nose of the plane. Without this banking, the plane would merely skid along sideways in the direction it was originally headed (provided that sufficient airspeed is maintained.) This same banking is used with the Rocket Belt when making a coordinated turn. The Rocket Belt's thrust tubes tilt instead of the wings. This is possible since the tubes are mounted on a gimbal.

To initiate this lateral "roll" motion, you need only tilt or roll your shoulders a small amount. The pilot clamps his armpit area tightly against the two control arms, and then gradually and gently raises one shoulder while lowering the other. Lower the shoulder on the side you want to turn towards.

During high-speed turns, centrifugal forces will cause your legs and lower body to swing outward with the turn, thus eliminating the need to "crank in" any hard sustained movements. In fact, a pilot must learn never to put in hard or violent control inputs because, due to action-reaction, what you get from violent control input is violent output!

Only expert pilots should ever attempt pushing an aircraft to the edge of its envelope. I have on occasion made such hard high-speed turns that my feet and legs actually swung through to be higher than my head. I

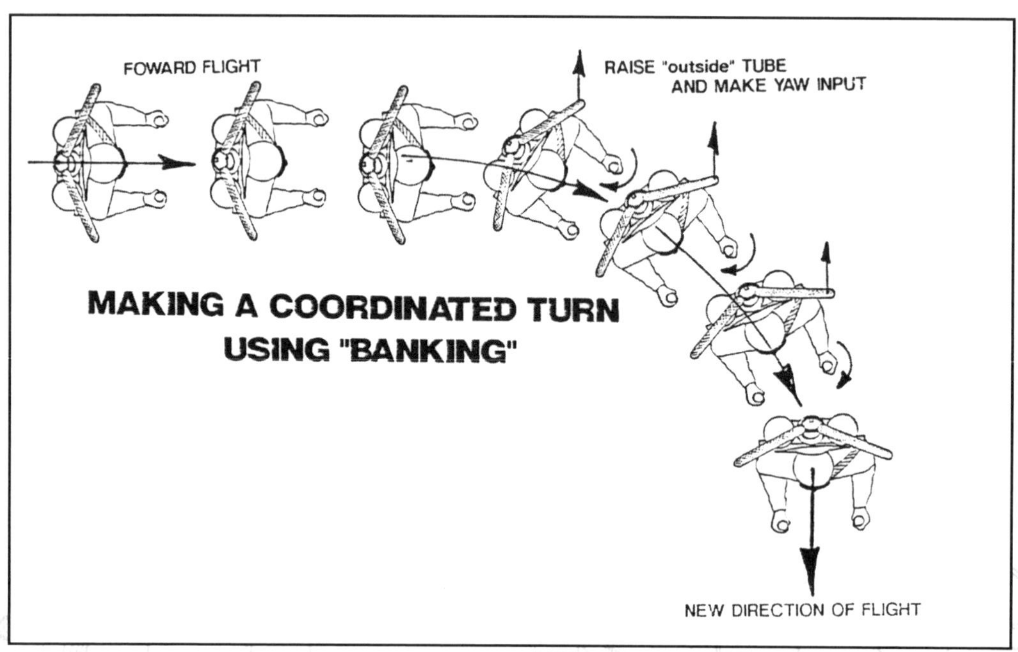

am certain that, had the Rocket Belt not been limited to 21 seconds of flight, a full roll could have been successfully executed.

Always keep in mind ... a Rocket Belt has no, zero, none, zilch, aerodynamic lifting surfaces (wings). The vertical thrust vector is the ONLY thing holding you up. A sustained hard turn will result in one thing: the flight of an anvil!

CHAPTER 13

Pre-Flight Checks

No responsible pilot ever gets into an aircraft without first going through certain pre-flight checks to ensure his own safety, as well as those under his flight path. Due to the Rocket Belt's complexity and the fact that there is absolutely no room (or time) for error, a mere kicking of the tires just won't do.

The following safety equipment is needed for pre-flight checks:

* Rubber gloves, apron and boots

* A pair of good safety goggles or full face shield

* A supply of clean running water, and hose with nozzle when available

First, make a visual inspection of each component. Look for any cracks or flaws in all welds and stress points on the fiberglass corset. Next, check all the straps and buckles and their mounting points. For the remaining checks, you have to use tools. A "wrench check" is done on every nut and bolt to make certain that they are tight. Screws are checked with the proper screwdriver. On fasteners with safety wires, see that all the wraps are intact and properly wound. Finally, refamiliarize yourself with the location of all the valves and gauges that were discussed earlier.

Before loading the H_2O_2 fuel into the Belt, you must make certain that there are no flaws in the system. A single small leak will not just shorten the flight time; this fuel is very dangerous and can cause chemical burns and fires. To check for flaws in the fuel system, you do pressure checks. This means pressurizing the fuel and nitrogen systems to their working pressures using NITROGEN ONLY. A nitrogen leak causes no hazards, yet it makes you aware of a problem with no harm done. SAFETY FIRST!

First, the nitrogen tank is filled with oil-free dry nitrogen. Never use compressed air or any compressed gas that is not compatible with H_2O_2 or that has been compressed (pumped) with an oil-lubricated compressor. There have been a few near-fatal instances when people who knew better used a scuba compressor as the easy and cheap way out. NEVER DEVIATE!

The nitrogen supply cylinders are filled to 6,000 psi (enough for six flights) and the bottles are always clean

Naval Reserve ship at Fort Niagara 1966. Note the angle of Bill's legs as he executes a hard banked turn around the ship.

and should have an inspection date stamped on them. The protective cap on a nitrogen supply cylinder is removed and the valve opened a very small amount. A short blast of nitrogen clears any dirt or moisture from the valve; be careful not to blow any debris into anyone's eyes. Safety glasses are necessary for everyone working around the Belt.

As you reach the 2,800-psi level, both fill valves are closed. The top of the nitrogen tank on the Belt is now hot, to the touch, because gases heat as they are compressed. Allow it to cool; as the tank temperature drops so does the nitrogen pressure. Once the tank has cooled, the valves are again opened slowly (Belt first) and the pressure is topped off. You have to be certain you have a full load of at least 2,500 psi, but don't exceed 2,800 psi. I always kept in mind that I was filling a tank designed for absolutely no more than 3,000 psi from a 6,000 psi source, and any inattention could have been fatal. It is very important not to be distracted from the procedure you are performing; keep your eyes on the pressure gauge on the Rocket Belt and do not exceed 2,800 psi.

Now remove the fill hose. First, close both valves tightly and then carefully crack open the connection where the hose attaches to the Belt to allow the remaining pressure to escape before attempting to detach it. Failure to do so can result in injury from the fill hose whipping about violently from the pressure.

Once the hose is disconnected, the fill valve dust caps should be replaced on the Belt and the supply cylinder. With the Belt tank filled it is now time to soap-check the high-pressure system for leaks. To soap-check you take a plastic squeeze-bottle filled with a mixture of Leak-Tek™ soap and distilled water and squirt it on all high-pressure fittings and joints. Any leaks will cause bubbles to form and pop. If any appear, tighten the joint with the appropriate wrench, should this not stop the leak, repairs are necessary.

Once the Belt's high-pressure nitrogen system checks out completely, proceed with fuel system pressure and soap checks. Remember, this is done with nitrogen only! Make certain the peroxide tanks are completely empty before beginning. Connect the drain hose to the fuel drain/fill valve, and put the open end into a clean pail half-filled with water. With ZERO pressurization to the fuel tanks, open the fuel fill/drain valve; any residual fuel will empty into the pail.

Now, check to see that the throttle is fully closed and that the safety locking pin is in place on the throttle handle. Next, slowly open the N_2 shutoff valve leading from the N_2 tank to the regulator. Then slowly open the fuel tank valve beneath the fuel pressure gauge on the right side of the Belt. You will be pressurizing the fuel system to a working pressure of 570 psi. Carefully watch the pressure build in the fuel tanks, being certain to stop and close the fuel tank valve should the pressure exceed 575 psi. The pressure should settle out to 570 psi.

If it does not reach this level, adjust the regulator to obtain 570 psi. Adjusting the fuel pressure is illustrated next. In the center of the regulator face is the pressure set screw (looks like a hole); towards the edge is the flow adjustment screw. An Allen wrench is needed for these adjustments. Insert the wrench into the center screw and slowly turn it counterclockwise, until a small flow or leaking noise is heard. To adjust the fuel pressure, turn the flow screw, being certain to watch the fuel tank pressure gauge and stop at 570 psi. These regulator adjustments are very sensitive and a light touch is definitely needed.

Once the fuel system is pressurized, soap-check every joint and fitting on the Belt's fuel system. As with the high-pressure side of the system, tighten or repair any fittings showing leaks.

Once the fuel and nitrogen systems are completely checked out, it is time to drain the fuel system. Only the fuel system is depressurized; the nitrogen system must be kept under pressure even when the Belt is in storage. This ensures that no outside contamination can make its way into the tank. It is neither necessary nor advisable to keep more than 500 psi in the system, but there always has to be a positive pressure on it.

To drain the nitrogen from the fuel system, first make certain the nitrogen supply valve leading to the regulator is fully closed. DO NOT vent the nitrogen through the catalyst bed via the throttle. Instead, SLOWLY push up on the fuel tank vent valve. Just in case there is any residual peroxide that might vent with the nitrogen, make certain no one is standing behind the Belt. You will hear a loud noise as the nitrogen escapes through the vent tube, which is located beneath the fuel tanks and pointed downward and to the rear. Once you no longer hear gas escaping and the fuel pressure gauge reads zero, close the vent valve. The Rocket Belt is now ready to fuel.

CHAPTER 14

Fueling the Belt

Before fueling, it is imperative to be aware of the dangers of handling 90% hydrogen peroxide. 90% hydrogen peroxide is an extremely powerful chemical oxidizer that can cause serious harm to those handling it or in the immediate vicinity. Only those personnel directly involved with fueling should be allowed in the area; others should be kept at a safe distance. At Bell, we always tried to do our fueling in a quiet, sheltered area free from wind or rain. I could tell stories into the wee hours of close calls while fueling with "90" (90% H_2O_2); take my word for it, close encounters of that kind are not good for the cardiovascular or nervous systems!

When fueling the Rocket Belt, personal safety is of paramount concern. Protective garments should always be worn and eye protection is a must. Dacron™ is a very good material for clothing; all the better flight suits are made of it. Any organic clothing should be avoided: Cotton, wool and some synthetics will burn very rapidly when they come in contact with 90% peroxide, as will human flesh! Put on all your protective gear now.

A generous supply of fresh potable water must always be on hand. Remember, water is the ONLY thing that can be used to combat peroxide accidents. When cleaning and flushing the fuel loading equipment, only distilled or deionized water should be used; any other water sources might contain contaminants that could cause a reaction.

The following additional equipment is needed to fuel the Belt:

* A hand pump with Tygon™ tubing for the peroxide
* A stainless steel standpipe for the fuel drum
* A small (two quart) stainless steel or aluminum pail
* Two plastic pails of clean potable water
* Five gallons of deionized or distilled water

The peroxide supply drum is made of pure aluminum and constructed like a Thermos™ bottle to afford the protection of a double-wall container. Normal temperatures pose no problems, but peroxide should be kept from freezing and away from extreme heat (above 110°F). The drum holds 30 gallons, enough for 6 flights; when making consecutive flights and not draining the Belt after each, it is possible to get 7 flights from a single drum. The fuel is very expensive, so using every last drop is important.

In the center of the drum's top surface is a cap. This is the port where the stainless steel standpipe is inserted. There is a second, smaller cap on the top of the drum, set off to the side; this is a vent to release pressure from the inner chamber surrounding the actual fuel container.

Before ever opening the drum, use distilled water to wash the area around the cap and flush off the top of the drum, being sure to wipe it down after flushing. A visual inspection ensures that there is no debris that could fall into the drum once the cap is removed. Should any get in, prepare for action. If it is organic, the dirt will react, and steam and smoke are very likely to vent in very short order. The drum itself is designed to rupture before any dangerous pressure buildup causes it to explode, but this causes fuel to run all over, setting things on fire. Your shoes are one of the first victims; always be prepared with lots of water.

A special wrench must be used to loosen the drum cap; don't completely remove it yet. The cap should sit over the opening until the moment you are ready to insert the standpipe. This reduces the chance of drum contamination.

Now, remove the standpipe from its case and wash it down inside and out with generous amounts of distilled water, then shake off any excess. Carefully set aside the drum cap with one hand and SLOWLY insert the standpipe into the drum with the other hand. Wait a short time to be certain that no foreign matter has gotten into the drum with the standpipe; it is much better to "launch" the standpipe than to screw it down and have the drum rupture from pressure. Once you are sure it's in and settled, listen for any hissing or bubbling noises. If there are any, wait a few more minutes for the contamination to decompose completely before proceeding. Now that everything is safely in place and you are reasonably sure that there isn't going to be any reaction within the fuel drum, tighten the standpipe.

Now the fueling hose can be hooked up. Take the hose from its case and remove the dust caps on each end; be sure not to drop them or get any dirt on them. Use a pan of distilled water to wash the hose thoroughly inside and

out, then shake out any excess and attach one end to the standpipe and tighten with a wrench. The peroxide is pumped into the Belt with a small hand pump, drawing the peroxide from the drum through the standpipe. You can now feed the hose into the rollers on the pump and remove the dust cap from the Belt's fuel fill valve, located at the bottom of the right fuel tank. Check to see that the valve is closed before removing the dust cap.

Before attaching the hose to the Belt, you first have to flush some peroxide through the hose. This is done by simply pumping about a cup into a pail that is half-filled with distilled water. Although the mixture in this pail will be far from 90% strength, it could still cause harm, so be careful not to spill any. Check the hose for any sign of reaction. If there are no bubbles forming, go ahead and attach the hose to the Belt and tighten it. Before opening the fill valve, it is first necessary to open the vent valve that is located on the back center of the Belt. Again, as with the fill hose, catch any overflow in a pan of water.

Place one end of the overflow hose on the overflow valve, and the other into the pail. Open the overflow valve and then the fuel fill valve; begin pumping the peroxide into the Belt slowly, keeping an eye open for a pressure buildup in the tanks, evidenced by a rapid release of gas from the overflow hose. After a minute of pumping with no problems, the pumping can be sped up to a reasonable rate. As fuel is pumped in, note the steady stream of bubbles coming from the overflow hose. This is the air from inside the fuel tanks being displaced by the incoming fuel. When the tanks are full, these bubbles will cease and fuel will begin flowing from the overflow hose.

Close the fill valve first, then the overflow valve; carefully drain any peroxide left in the overflow hose into the pan. Carefully empty the liquid into the pail of hot water, (meaning that pail of water that the cup of purge peroxide was pumped into during fueling.) Next, disconnect the fill hose from the Belt, flush the hose, replace the dust caps, and then store it.

If this is the only flight to be made for several hours, it is necessary to drain the fill hose, remove the standpipe, flush both with distilled water, store in their containers, and cap the drum. Finally, take the pail of "hot" water to a drain that is, free of grease and oil and, with lots of flowing water, slowly pour out the mixture; keep the water flowing for several minutes after. The Rocket Belt is now fueled and flight-ready.

Another early test on a tethered nitrogen rig in December 1957

CHAPTER 15

The Tether System

There is no more important piece of training equipment than the tether. The tether system is our safety net, and was my saving grace on a number of occasions before I learned to solo and fly free.

It requires two people to operate properly: a catcher and a trolleyman. The catcher must be physically large, in excellent health, strong, and weigh more than 215 pounds; the trolleyman need not be so big. Both men must be quick on their feet.

The catcher is the most important person in the training program: he ensures not only the well-being of the pilot, but also that of the Belt. The catcher's job is to do just that "catch" the pilot when he gets out of control and safely lower him to the floor.

The tether itself is a small steel cable on a pulley system. A tether bracket with a pulley attaches to the Belt; the cable passes through it and runs up through another pulley on the trolley high overhead on an I-beam. The cable ends at the trolley and at the catcher's harness. The trolley arrangement hangs from the I-beam on a set of free-moving rollers. On the bottom of the trolley are a pulley that the tether cable passes through and two steel rings: one that the tether cable mounts to and another that the trolleyman's rope attaches to. A good catcher could be a Defensive End for the Buffalo Bills!

Buffalo Bills: large, fast on his feet, and used to being beat up in the line of duty! The job is a thankless one, and anyone chosen for the job has to be there by choice - his own!

On more than one occasion during the early days of training at Bell Aerospace, catchers actually received whiplash (and even a few chipped teeth) while saving the necks of the pilots who weren't quite ready for prime time. During my training (I was a difficult student), I can recall seeing Ernst Kreutinger and Ed Gaiser, both catchers at the time, actually flying UP to meet me on my way DOWN. Them on one end of the tether cable, me on the other, but I was the only one who was supposed to leave the hangar floor! I owe my life to the late Ernst Kreutinger. He was the catcher on the day I was training and a crucial weld failed. I had the ride of my life! Ernst never once lost his cool; and Doug

"THE CATCHER"

PILOT

"THE TROLLEYMAN"

THE TETHER SYSTEM

Meiklejohn did a great job on the trolley. The two of them exemplified the meaning of teamwork.

Not to be overlooked or undersold is the trolleyman. His job is to keep the tether trolley directly over the pilot. This lessens the chance for mistakes by the catcher. If the trolley "lags" on forward flight, it could tighten the cable, tending to hold the pilot back. If the trolley "leads," this tends to pull the pilot forward. Remember, in this almost weightless situation, it takes very little physical input to move the Rocket Belt in any direction. The trolleyman is not concerned with "in and out" as the catcher is, because the length of his rope remains constant at all times. However, it is very important for him to anticipate the next move and try to stay with the action, not ahead nor behind.

The tether crew has to wear good eye protection; the blast causes the smallest foreign object to be a dangerous projectile. They also have to be sure that there are no grease or oil or wet spots that could cause them to slip and fall. Both tether men should wear a good pair of leather work gloves to protect their hands from the cable and rope, and to grab for the hot Belt in case it swings wildly out of control.

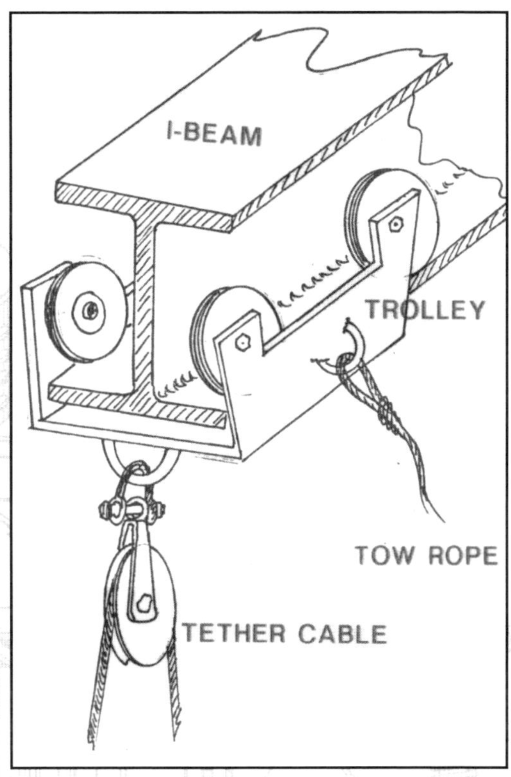

During the training phase of the program, not only do the catcher and the pilot get beat up, but also the Rocket Belt Control arms, thrust tube components, and welds have to be inspected thoroughly before and after each flight, keeping a sharp eye out for any cracks, bends, or misalignment of parts.

The catcher probably does twice as much running as the trolleyman, because he must run in and out as well as back and forth. As the pilot begins to lift off the floor, the cable slackens. At no time can the catcher allow enough slack that a loop or a large dip could

form and get tangled around the pilot or Belt. This is a deadly situation and must be constantly guarded against. If a slack cable wrapped around the pilot's neck, the pilot has only as long as the fuel supply lasts to enjoy life as he has known it to that point. During one training session at Bell, a pilot was very nearly castrated when a cable loop wrapped around his upper leg and he cut power. With this in mind, the catcher must fully realize the importance of his job. It is much more than to supply ballast!

As the pilot lifts off, the catcher must back away from the pilot to take up slack in the cable. Then, as the pilot begins to move forward (or, if early in training, backward), the catcher must also move in that direction. While the pilot goes up and down, back and forth, the catcher has to keep on his toes and follow every move. At all times, the catcher must allow only enough slack in the cable so as to not interfere with the pilot's freedom of movement. Keep in mind that the pilot is almost weightless. The slightest interference from the cable could cause control problems, and the pilot might not even realize that the difficulty was from an outside source: the catcher.

While we were learning to fly the Rocket Belt, our minds were so focused on controlling the flight that we needed all the help we could get from every member of the team. The catcher always has to be ready to reel in the trainee pilot, and must never become distracted from the chore at hand. This is a deadly serious business and that fact cannot be forgotten. All persons not directly related to the training should be kept out of the area. This obviously reduces the risk of injury to the visitors, but it also reduces the number of distractions the pilot, catcher and trolleyman have to deal with.

CHAPTER 16

Tethered Flights

It's show time! Transport the Belt to the training area and put on your flight gear, basketball-type knee guards, several layers of Ace™ bandages wrapped on your lower legs (to protect against steam burns), and a Dacron™ flight suit.

The Belt and stand are fully prepared for flight and in position directly under the tether trolley at one end of the training beam. Attach the tether bracket to the Belt; it bolts onto the Belt's control arms where the arms join the thrust tubes. The tether bracket is high enough to reach above the pilot's head. With the catcher and trolleyman steadying the Rocket Belt, the cable and pulley are attached, and you strap in.

Tethered Pre-Flight Check List

* Throttle locked and safety pin in place
* Nitrogen supply at 2,500 psi
* Fuel pressure valve closed
* Open nitrogen supply valve to the regulator
* Slowly open the fuel pressurization valve, and adjust to desired fuel pressure

The catcher gets into his harness, attaches the tether cable to it, and then backs away until the cable is tight. At this point, he should look the system over to make sure all is in order, with no crossed cables or fouled pulleys. If everyone is ready, the trolleyman removes the pins from the stand's stabilizer arms. The catcher now backs up to help lift you and the Belt away from the stand. After you become accustomed to the routine, you won't need the help, but early in training it is a good idea.

To get a feel for the weight, the catcher now lifts you 6 to 8 inches off the floor. Should anyone's harness need adjustments, this is the time to make them. Before proceeding, the crew checks the area once more for any hazards. The stand should now be moved to a safe distance, but kept handy.

The first few tether flights are limited to about 60% fuel pressure, insufficient power for takeoff; however, it is enough power to push you about and give a good feel for the control inputs needed. If you tried with full pressure, you would be thrashed around without really getting much feel. If the pressure

NASA Test Pilot Lee Persons tries out the Bell Pogo on the tether-1966

climbs above 340 psi during pressurization, the excess pressure must be bled off at the regulator. To do this, a technician inserts a wrench into the regulator and adjusts the pressure to the desired level. At this reduced fuel pressure, there will not be sufficient power to lift you off the floor.

Now, standing at one end of the tether training area, the trolleyman checks that the trolley is directly overhead. Look to each man for a nod that they are ready and have their ear protection on. Remove the safety pin from the throttle handle and firmly (but not to the point of white knuckles) grasp the two control handles.

Either the trolleyman or the catcher will signal when the thrust tubes point straight up and down. Looking straight

ahead, SLOWLY open the throttle. The control arms immediately jump from the built-in upward control force. Try translating forward. The reaction forces will shove you along the ground, so walk in the direction they push you. Make several starts and stops. As fuel is used, the Belt becomes lighter. As the Belt gets lighter, you will actually be on your tippy toes and very light. Due to the lower fuel pressure, the run time will be longer than 21 seconds.

Once the first load of fuel is exhausted, the Belt is put back on the service stand, and you, the catcher, and the trolleyman have a little "skull session." Your observations and feelings are tossed around, and notes are compared. Get in the habit of logging all facts and personal comments pertaining to each flight.

Refuel the Belt and make another run like the first. On your second run, try some lateral control and yaw movements; again after the flight, you should compare notes and refuel.

For the third flight, the same procedure is followed but the fuel pressure is increased to 70%. This will be noticeably more powerful and everyone has to be prepared. Actual lift-off will be possible towards the halfway point because of fuel burn. You will barely be flying on this first "departure" from Earth, but when your feet leave the floor for the first time, the sensation is almost overwhelming. Once this milestone is passed, it gets easier to concentrate on what you are doing. Remember though, the increased pressure shortens the flight duration. If there is sufficient time, make several 5- or 6-foot long hops, then refuel.

For the next flight, fuel pressure is increased to 90%. This is enough to get airborne almost immediately; be ready for it, and also for the increased upward force on the control arms. Don't attempt anything more than short, low-level hops to practice landing control. Learn to walk before you run. As fuel burns off, the throttle becomes much more responsive and it is very easy to over-control. At this fuel pressure, the 21-second limitation now comes into play.

Continue to practice at the 90% power level until you feel comfortable and can continually take off, translate a short distance forward, and land without running. Then try adding a 180° turn so you can return to the takeoff point to land.

As you practice, you will reach a saturation point in your daily training. There will be times when the tether fouls your flight. There will be times when you don't even know what fouled your flight. It was our practice not to exceed 3 flights per day. If we were not progressing, we would cut back on the workload and ease up a bit. All of this may seem rather straightforward and simple, but don't be fooled by it. It is not simple and no one learns it overnight. Regardless, the ultimate success rate of our training was good.

Once you begin to feel in control, you become more aware of the buzzer and can begin to gauge your flight by it. Before you can move outside for your first free (untethered) solo flight, you must be fully aware of the buzzer and fully in control of the machine. Try a few flights with full pressure to build your confidence; if you do them consistently without aid from the tether, it is time to solo.

CHAPTER 17

Free Flight

Pick a smooth grassy area, free of obstructions. The temperature has to be above 50°F; the exhaust condenses below that temperature, and the steam will obscure your vision. Without being able to see the ground, you have no altimeter. Observers should be kept to an absolute minimum for the first few free flights.

Bill Suitor, Disneyland 1965
"I love the smell of peroxide in the morning!"

For my first free flights, the fuel pressure was kept low and I just tried to relax and treat it the same as the last successful tether flights. Nothing much was expected, because this was not the time to stretch my ability or my neck. Successfully and safely taking off and landing was a great accomplishment. Do you realize how few people have ever done just that? In fact, as of 2009, more people have walked on the moon than have free-flown rocket belts.

On my first flights, I took off to an altitude of 4 or 5 feet, then moved slowly forward until I heard the first buzz; at that point, I made a slow controlled landing. A tumble from that height didn't hurt much more than my dignity. Everyone to date had made some sit-down landings, so it was no big deal to "join the club." I picked myself up, brushed off and started over. After a tumble, I liked to brush up on the tether before the next free flight. This checked out both me and the Belt. Whenever there are doubts or problems, the tether is the place to get the answers. It is the only safe place to try things out.

If you have successfully completed several simple free flights like this, now attempt a few turns. Place an object under the flight path, 50 to 75 feet away. Take off and fly toward it; as you approach, turn away from it, either

A NASA student takes a tumble and joins the club during free flight attempted landing with Pogo - 1966.

to the left or right, whichever is most comfortable. Remember two very important things: fly by the buzzer, and keep it low and slow. "There are many old pilots and there are many bold pilots, but there are no old bold pilots!"

You could then try some 180's. Fly out and around an object and return to the takeoff point to land. Then do several turns to the left, then several to the right, always listening closely for the buzzer. A slalom course is a good way to practice turns without getting into trouble. The "pylons" should be such that they won't harm you if you fall on one.

As your self-confidence grows, try stretching the distance a little more and start flying over obstacles like cars and trucks. The next goal is to fly over small trees. Altitude practice should not be rushed; close attention must be given to the way the descent rate is set up. Remember, you have to be able to walk away from each landing.

Take extra caution not to get wrapped up in this experience and forget about the buzzer. The sensation you feel as you make these new flights of freedom will overwhelm you. Believe me, I know. I have made around 1,200 flights and I still get a tremendous thrill each time I soar into the sky. The most thrilling thing for me is to see my shadow racing along on the ground below me. It is a sensation unlike any other that you will ever experience.

As I became accustomed to flying, there was a time when I could just strap and fly away as naturally as I walked. It was then I realized I had "made it." From that point on, the flights were very enjoyable. However, I never lost the butterflies before each flight. If you feel you have, you are in big trouble. You must always respect the Rocket

CHAPTER 17 Free Flight

Bill Suitor, Washington DC, 1967

Belt. It is extremely dangerous.

All Rocket Belt pilots request more power as they become more proficient, and have to be reminded that "what goes up, must come down." We must never get carried away with the desire to go higher and faster, because we may end up doing just that: being carried away! A "hot" throttle is every Rocket Belt pilot's dream come true. There is no other feeling on Earth like hitting that throttle and feeling like you have left your sneakers behind. Or the sensation of dropping like a stone, hitting the throttle and feeling the "brakes" grab. These sensations and thrills are unlike any other. However, never let them run away with your better judgment. I am living proof of that!

Bill Suitor, Disneyland,1965

CHAPTER 17 Free Flight

Bill Suitor, as James Bond in "Thunderball", Paris, 1965

Bill Suitor, Los Angeles,1965

CHAPTER 18

Fine-Tuning Your Flights

Okay, here you are: a free-flying, proficient Rocket Belt pilot. Now is the time to fine-tune your ability. Time to learn a few tricks of the trade.

After the Rocket Belt proved that man could fly without wings, it became nothing more than a Hollywood curiosity. Given the expense of each load of fuel, I could not very well do much practicing. Knowing that, each flight I made was both for show and for practice, whether for a movie, a television show, a football game or a ceremony. I just had to make it look like every second was not really practice. It was all show biz, so I had to give them their money's worth every time.

There are limits to the flight controls of the Rocket Belt but there is no limit to our imaginations. As I flew, I tried to be aware of each and every thing I did, what input caused what output; every precious second of flight had to be a learning experience. By sticking to this practice, I learned that there are countless things I could do to improve my flying by using my brain and some "body english" to execute maneuvers.

Do you realize that you can inadvertently initiate forward movement merely by tipping your head forward, or moving your feet forward? You can, and that is because the center of gravity (CG) changes. The next time you see photos (or better yet, movies) of the Rocket Belt in flight, note the pilot's legs and their position, They tell you a wealth of information about a pilot's proficiency.

Greenhorns invariably use their legs for control. When I see a pilot pumping his feet up and down or back and forth, this tells me that he's chang-

Three of the Bell "B" Belts

ing his CG and probably isn't even aware of it. These changes in a pilot's attitude will tell you whether he has it or not, or maybe that he never will have it. If you need to change your CG to control your flight, then you are not using the flight controls properly. You are probably too busy worrying about staying alive, and you really don't know what you are doing yet. Get back on the tether to practice and learn more. Problems can also be caused by being improperly strapped into the Belt; you are not properly aligned with the control arms, or your weight might be hanging on them.

It is important to keep your feet together and your legs pointing straight down at all times. During my training, soon after free flight, I picked up the bad habit of lifting my feet forward as I flew. My instructor, Ernst, made me fly while holding a chunk of two-by-four wood between my feet. I had to concentrate on my feet, and this corrected the problem.

You will note in some pictures that the pilot seems to be leaning one way or another, yet still aligned through his CG. The pivot point (gimbal) for the flight controls is centered directly behind the pilot's head. When he pushes down on the control arms, he gets a forward push at that point and his upper body moves slightly before his lower body. His feet lag, giving the angulation. During a real "barn burner" (as I like to call high-speed flights), this also happens. Due to the combination of the push up-high, and wind resistance down-low on the feet, the feet really lag behind. This will be quite evident in the opposite direction when the pilot puts on the brakes. During a fast stop, the reverse thrust input is centered high on the mass, so the lower body tends to swing forward past the head. This only lasts a moment.

During any angulation movement, be aware that as your legs and lower body swing one way or another, the thrust vector is becoming more horizontal; this steals vertical thrust. Lateral angulation is a double-edged sword. Use angular "stall" to your advantage to lose altitude when you are too high. Use it to your disadvantage by making too hard a turn and unintentionally falling out of the sky.

Great caution must be exercised in a hard fast turn. As you make a coordinated turn, the forces swing your legs to the side. The harder the turn, the higher the swing. On several of my flights that were very high (more than 100 feet) and very fast (exceeding 60 mph), I actually had my feet swing through so that they were almost high-

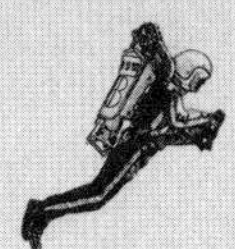

er than my head! In this situation, all vertical thrust is lost, and you become "the anvil." Be ready to get yourself out of it - quickly! This maneuver is meant neither for the inexperienced pilot nor the faint of heart. As you make like an anvil up there, your guts find their way up into your mouth. You must swallow hard, hang on, and (as always) do not forget to pay close attention to the buzzer. There is no lonelier feeling than to be very high, going very fast and realize that there are only 6 or 7 seconds of flight time remaining. BELIEVE ME!!

Whenever you set up a rapid rate of descent, you always need sufficient power to "stop the drop." A good rule to follow is that if power is less than optimum, you fly a conservative flight. You'll sense a less-than-optimum power during a sluggish takeoff and climb, or by "falling out" as you start to translate forward. This can be caused by a number of factors. The fuel pressure may be set too low; the catalyst pack may be failing; cameras or other equipment may have increased payload; or, the pilot may have consumed too many hamburgers before the show!

You could find yourself in a situation with insufficient power to stop a descent and hit with the throttle wide open. Don't panic! It's not as bad, as it sounds. I had it happen on a number of occasions. If you are lucky (and I have been very lucky), your descent will be slowed enough by the full-power setting, and as you hit, the 330 pounds of thrust blows you right back into the air without so much as a knee flex. Stay sharp as a tack and do not lose your cool when this happens.

During training and practice, you concentrated on cutting power the instant your toes made ground contact. Otherwise, you bounce like a ball and were blown back into the air. However, in this situation, cutting the power is the WORST thing you can do! There you are, a foot or two off the ground with the heavy machine on your back and WHAM, over you go, flat on your back!

Without power, there is no-control; as long as the machine is running, you can maintain or regain control. If you find yourself bounced back into the air due to a hard landing, DO NOT CUT THE THROTTLE! Keep control, steady yourself and land again. Although this sounds like it takes many seconds, in reality it only takes a couple. And, you only weigh a few pounds when you hit under full power; your feet may sting a little, but at least you can walk away from the flight.

Now imagine you are flying the Rocket Belt. I will set up an imaginary situation, and then share my flight experience to help get you back on the ground.

You are flying along at 105 feet altitude and after 13 seconds, have traveled out 280 feet. Now, you must turn 180° and return to the takeoff point. Turning slowly will not only take too much time, but you would still be over 100 feet above the ground. Losing altitude had better be at the top of the agenda as you only have 8 seconds left. However, if you reduce power to lose altitude, you will never make it back to the landing site in time. What to do?

Keep the throttle wide open and crank in a hard 180° turn with lateral tube control, yaw control and a certain amount of body english. This accomplishes several things:

* It minimizes your turning, radius: less distance to travel.
* The Belt will "stall and fall." When combined with the forward tube deflection present before you initiated this-turn, the strong lateral deflection will steal so much vertical thrust that you'll drop like a rock.
* By maintaining your speed, you'll be closer to home by the time you recover from the turn.

As you straighten out the turn, the lateral thrust is restored to vertical, and your descent rate is slowed. Now, halfway home, you are heading in (forward and down); at this point, pull back on the controls to begin braking. This steals vertical thrust as your legs swing forward, and you still have throttle left. As your legs swing through, you will decelerate very rapidly. With some practice, you can make this look like one smooth continuous move, as if "you actually know what you are doing." If done perfectly, your feet touch down at the same moment that forward travel stops. They had better!

You'll have to feel your way through the body-english part of practice. No two people are alike and no one interprets anything the same. There are no set rules that can be taught, such as "X" amount of this and "Y" amount of that. You have to learn from every move you make. Even after all the hundreds of flights I made, I still learned or experienced something new each time I flew. The most important thing to remember is that there is no room for error. Anything you try needs to be thought out thoroughly long before you ever leave the ground.

These basic instructions are enough to get a pilot up and get him there, but to become an expert at flying the Rocket Belt, you have to become the "Man-Rocket." You and the machine must become one. You will need to fly merely by doing rather than by thinking about the controls, just as when you run, you don't think "now I must put my left foot in front of my right."

CHAPTER 19

A Nice Demo Flight

My first demo was in July 1964 for some Apollo astronauts. From left: Ted Freeman, Bill Suitor, and Gene Cernan (the last man on the moon)

Once you master the Rocket Belt and feel comfortable in front of a crowd, here are a few things you can do for a nice demonstration flight. This type of flight certainly impresses the non-flying public's untrained eye, but it really knocks the socks off an observer who appreciates the difficulty of flying a pure thrust vehicle with so much raw power. Many aerospace experts, including astronauts and test pilots, have left a good demo shaking their heads, amazed at the fine control and maneuverability of the Rocket Belt. People don't realize just how long and hard the training was to reach that degree of flight prowess.

Filming the "Here Comes Tomorrow" TV show in 1971.
From left: John Glenn, Nelson Tyler, and Bill Suitor

For this demo flight, you will show stability, speed, rotation about the vertical axis, forward, reverse and lateral control, as well as a coordinated turn. You need an area of at least 600 feet square (the larger the better for good crowd control), free of dirt and stones that could be blown out and injure someone.

Before the flight, an announcer gives a short history of the Rocket Belt's development and a description of what you are about to do. Remember, the Belt is very loud; tell people to cover their ears, especially small children. Be especially cautious if there are pets present, as they might panic. Don't laugh; it would not be funny to be an amusement park guest and have Samu the elephant decide to exit posthaste out of the area of the flight!

Show time! Places everyone. Standing in the center of your "launch area," pressurize and make your final checks. Double-check that a supply of water is close by, just in case; SAFE-

TY FIRST! When the announcer is done, warn everyone again to cover their ears. Give a short throttle blast to check out the system; better to find a problem before you take off rather than after. Recheck your pressures and then turn to face the crowd. At any demonstration with kids present, have them take an active part in your flight by doing a countdown after the warm-up blast. They love it! 5-4-3-2-1 ...

Do a vertical lift-off and, as you rise, put in a yaw command for a slow 540° rotation (one-and-a-half times around). Now you are facing away from the crowd, with about 40 or 50 feet of altitude. Then; "dump it," "hit it," "kick it," but in some similar manner, get it to high speed in a hurry. With the throttle fully open, travel out a couple hundred feet. Now begin a nice coordinated turn, try not to fall out of it, stay in control (no skidding or sliding) and come directly back to the takeoff point. Descend as you approach, timing it so you stop about 3 feet off the ground; while hovering, side-slide to the left or right a few feet. If there is still enough time left, you can really dazzle them by touching one toe to the ground and doing a ballerina-type pirouette. After landing, by all means, salute the crowd.

Trust me, this demo flight will send them home talking to themselves. Especially the experts. "Impossible, impossible," they'll mutter, but seeing is believing

After the flight, plan on a question-and-answer session, because there will be lots of questions. Stay up-to-date on all aspects of the Belt's operation so you can answer questions simply and with confidence. Make sure people approach the Belt from the front to make sure they don't accidentally touch the red-hot thrust tubes. Oh yes ... only after the pressure has been relieved from the fuel tanks can people come close enough to kiss you!

Peter Kedzierski - Paris Air Show - 1963. Note Wally Schirra's Sigma 7 spacecraft at left.

Conclusion

I hope that you come away from this both informed and entertained. The Rocket Belt is fun, as well as serious business. I have spent decades on the project since I joined Bell Aerospace in 1964.

One day, travel with flying belts will be common. Not that everybody will have one, but they should not remain an aviation curiosity forever. I don't think a "Buck Rogers" backpack will be the answer, but some sort of "Individual Mobility Device" (as the military used to call it) will be developed. So far, every attempt has been doomed by either bad project management, lack of funding, or a lack of a clear purpose.

From this point onward, you will find yourself longing for more flight time as you realize the "boundless halls of air" that await. That 21 seconds becomes a real bummer. Study, dream and promote the design and development of new and better flying machines. It is up to you, the young people of this country, to build upon our experience and make it work. The last dreamer died the day John F. Kennedy was shot. What we need is some dynamic leadership to take us from here. The future belongs to you.

In closing, allow me to emphasize as strongly as I can ... NEVER attempt to build your own rockets. This book is in no way intended to be a guideline to attempt such a thing. Only highly-trained aerospace experts working in controlled conditions can do this safely; finishing college will help prepare you for this type of job.

Thank you, and good luck in the future!

William P. Suitor

Preparing for our entrance at Superbowl #1. From left, Ed Ganzcak, Charles Kriener, R. Courter, Ed Gaiser, Bill Suitor, and Ed Mullens - (Below) Entrance at Superbowl #1. Courter at left and Suitor at right

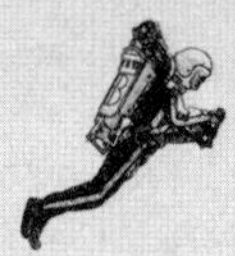

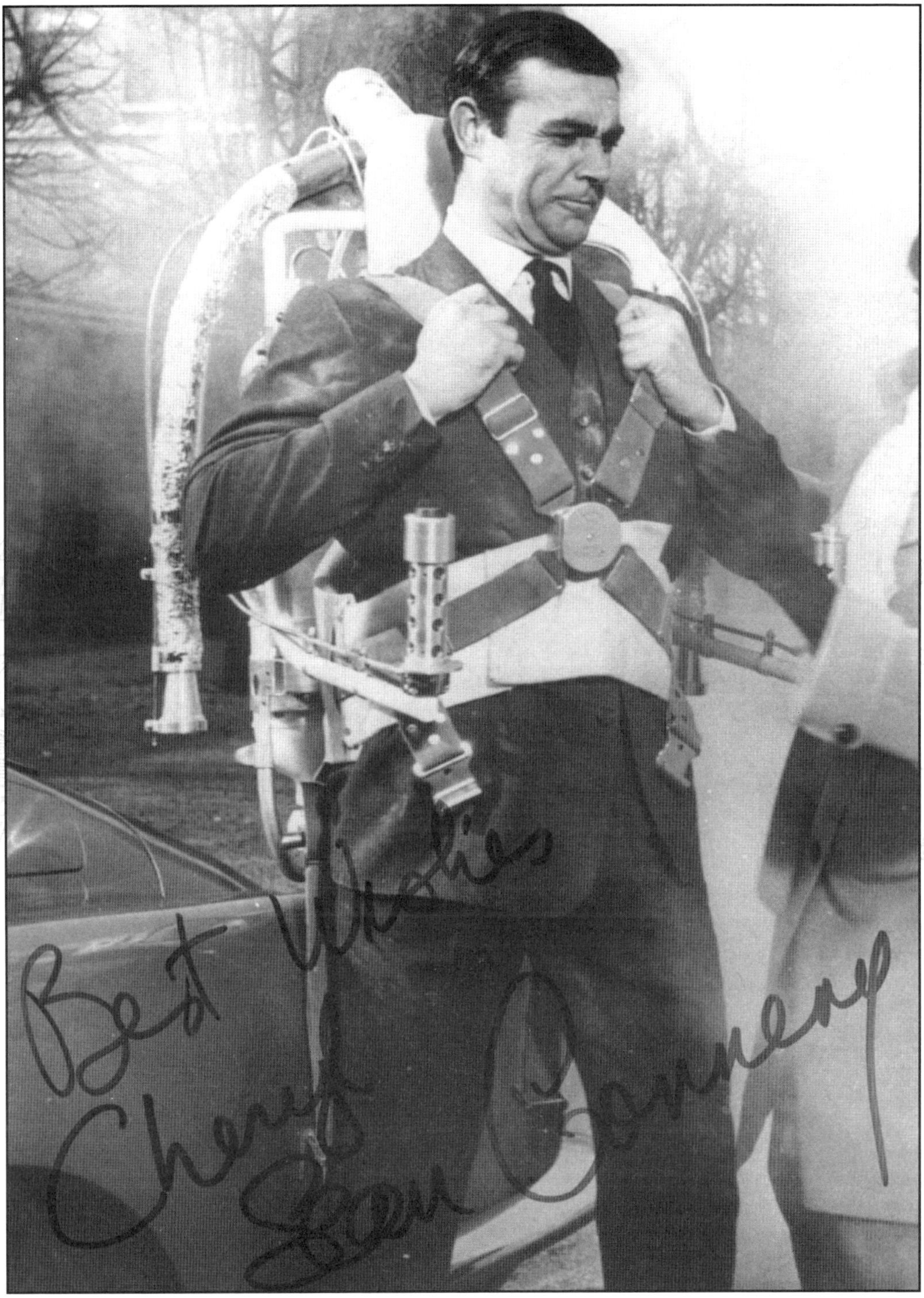

Sean Connery kindly signed pictures like this one from "Thunderball" for the Suitor family.

©Danjaq - United Artists

The Bell Rocket Belt team. Suitor at left. Johannesburg, South Africa 1965

First free flight of the flying chair. L to R , Dr. Roland Walker, "Mitch The Fireman", Ernie Kreutinger, Doug Meiklejohn, Gordon Yeager, Bill Burns, Ed Gaiser, Wendell Moore, Les Beukima, Bill Suitor, Bob Courter seated

Dual demonstration flight. R. Courter left, P. Kedzierski right - 1963

The Mall, Washington DC, 1967

Bill Suitor preparing to make Army training film, 1966

Navigating the Niagara rapids.

Bill Suitor with Walt Disney. Note the TWA moonliner in the background.

Bill Suitor on the set of "The Fall Guy" with Lee Majors, 1982.

Posing for Dean Conger of National Geographic with Wendell Moore. (above)
Buffalo Bills vs San Diego, 1966. (below)

NASA Langley, inspecting 1/6th G Lunar trainer
L to R Bill Suitor, Doug Meiklejohn and Edwin Satterle - 1967

Bill Suitor with Neil Armstrong meeting again after almost 40 years.

photo by Berry DiGregorio

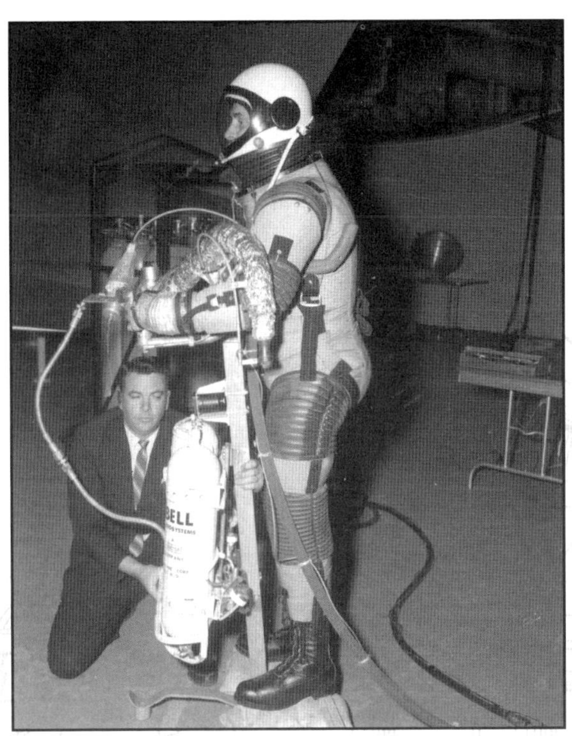

POGO being tested with early pressure suit. Early 1960s.

Wendell Moore with prototype of Bell Jet-Pack

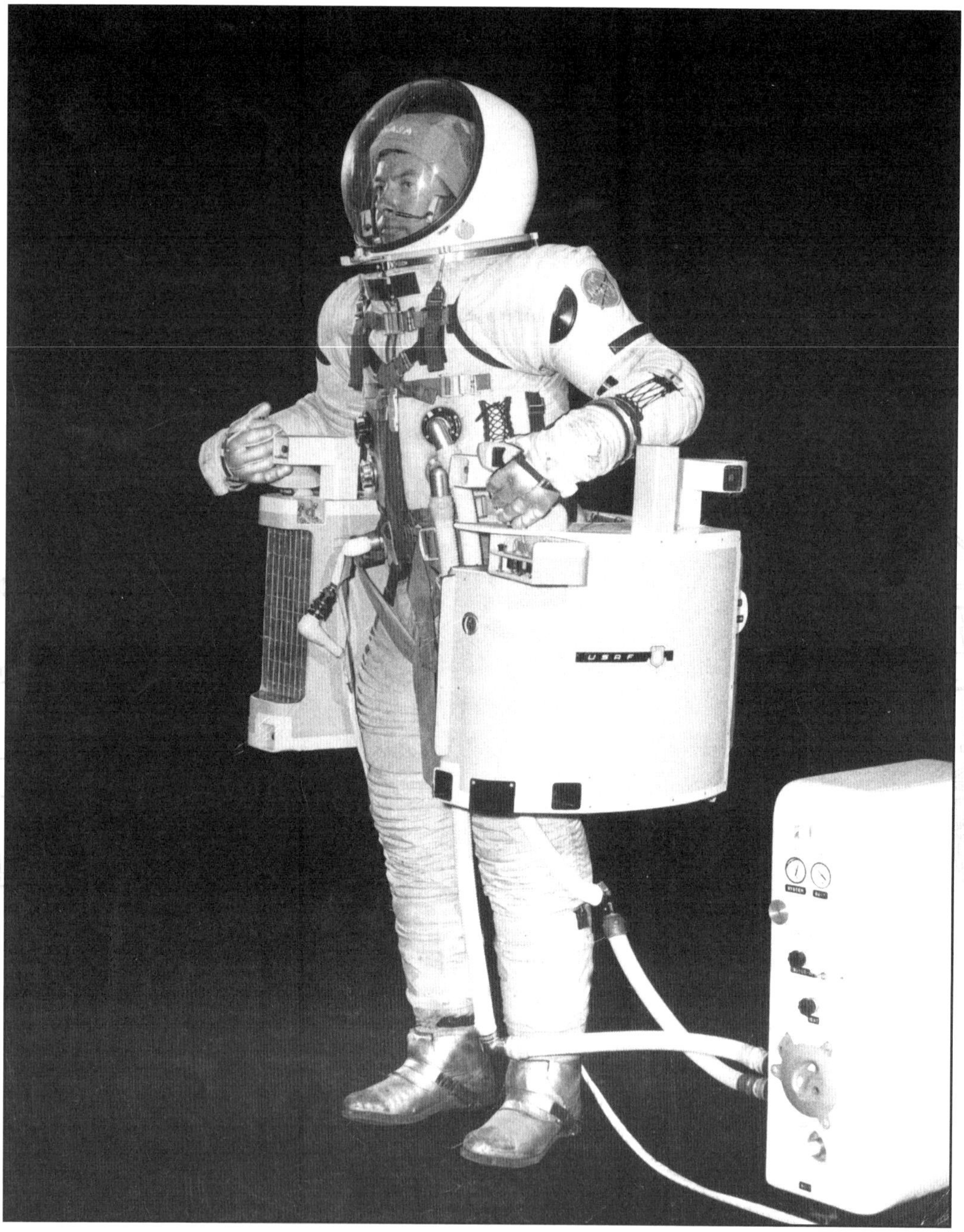

Mock up of AMU (Astronaut Maneuvering Unit) on Apollo "Soft Suit", this was the "Next Generation" of the Bell Zero G Belt.

Lunar Pogo at NASA Langley, Virginia, May 1967 flying on the giant 1/6th G training rig used by the Astronauts to practice 1/6th G landings

Bell Lunar Pogo doing tether tests in the Bell hangar wearing the Apollo Soft Suit. The large vacuum cylinder used to lift 5/6th of the weight, thus providing 1/6th gravity, can be seen overhead.

Initial test rig for 1/6th G Lunar Pogo, Bell Hangar, 1966.

Initial test rig for 1/6th G Lunar Pogo, the weights were used to dampen pitch and roll to what would be encountered in Lunar Gravity. Bell Hangar 1966.

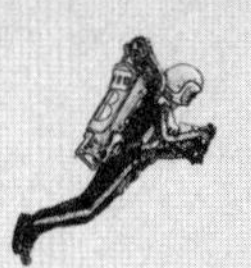

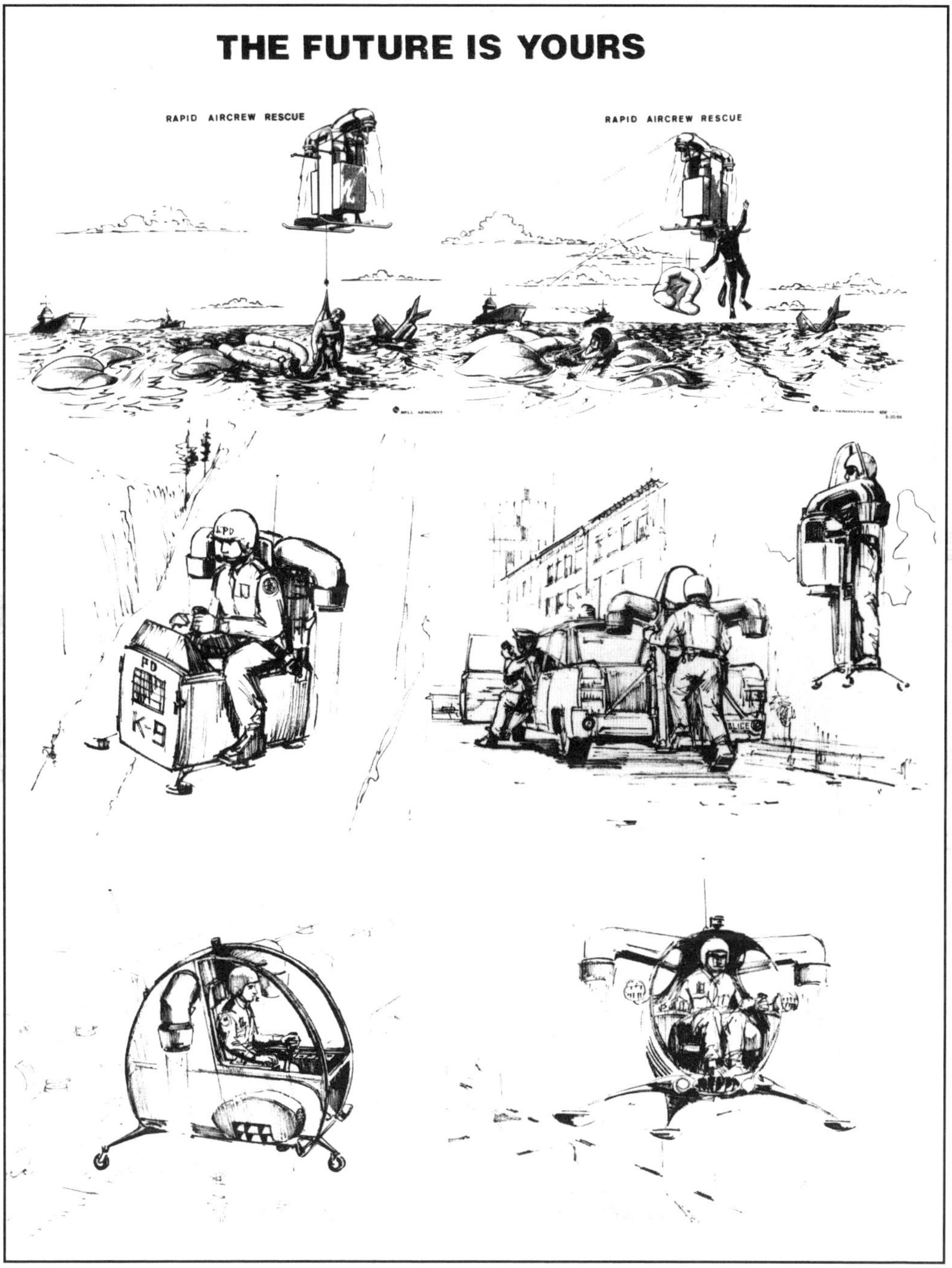
THE FUTURE IS YOURS
RAPID AIRCREW RESCUE
RAPID AIRCREW RESCUE
LPD
PD
K-9

Bell proposed JET POWERED flying machines based on the successful tests with the Rocket Powered Bell "POGO" and Rocket Powered "Flying Chair", designed by Wendell Moore and John Hulbert.

Bill Suitor tests the Reverse POGO on the tether in the Bell Hangar, 1966.

Yaege and Burns flying the Dual POGO at Bell, June 1967 (right)

The Bell Rocket Belt patents begin with a small rocket developed to control the Bell X-1A supersonic plane. The control rockets were built by Wendell Moore and Kurt Stehling. Two patents were issued for the H_2O_2 reaction controls.

Bell then built the reaction control system for the Mercury spacecraft. MacDonald Sill came up with the new gold-and-silver catalyst bed to speed up the response time. These rockets were used in the X-15 aircraft.

Richard Deegan invented a device for testing the motors thrust more accurately.

Arthur Gorbaty invented a Proportional Reaction Control Rocket in 1960 to accurately vary the amount of thrust and position the vehicle in flight.

Using the H_2O_2 system Wendell Moore built the first man-carrying rocket belt in 1958. He tested it using a nitrogen rig.

During the gas rig testing a corset device was developed to support the rocketbelt and fuel tanks. This became known as the Hip Pack and led to most of today's heavy duty mountaineering equipment.

In 1959 Wendell Moore developed a zero gravity belt for use in space. It was tested by the US Air Force in Ohio. To come as close to zero gravity as possible Wendell Moore tested it using scuba gear in a swimming pool.

Bell's John Chaplin took the basics of the rocket system and invented a stabilizer for air-cushion vehicles (hovercraft).

Wendell Moore designed a lightweight pump arrangement to increase the rocketbelt's range, and even tried using it attached to a Rogallo-type flying wing.

John Hulbert took the hydrogen peroxide system and used it to drive a turbofan.

For use underwater, Mr Hulbert devised a Hydrodynamic Propulsion System.

Hulbert and Frank Bond then refined this system by using a closed-cycle steam turbine which could provide thrust, heat and electrical power to the operator. The decomposed H_2O_2 also supplied oxygen.

Moore and Hulbert teamed up to design a Turbo-Jet Flying Belt. The US Army and ARPA awarded Bell a $2 million contract for its development. It used hydro-carbon fuel and was expected to fly easily to several thousand feet altitude.

Mr Hulbert then added a simplified auto-pilot.

Moore and Edward Ganczak took the propulsion system from the rocketbelt and created the POGO stick flying machine. A stand-up vehicle which allowed the operator to simply step aboard and fly away.

Moore and Ganczak then adopted this system to a flying chair concept. Both were expected to be able to carry an operator and 150 pounds of equipment on the moon on round trips up to 12 miles.

Nov. 14, 1961 W. F. MOORE ET AL 3,008,672

ALTITUDE RESPONSIVE AIRCRAFT JET CONTROL

Filed Oct. 4, 1960 2 Sheets-Sheet 1

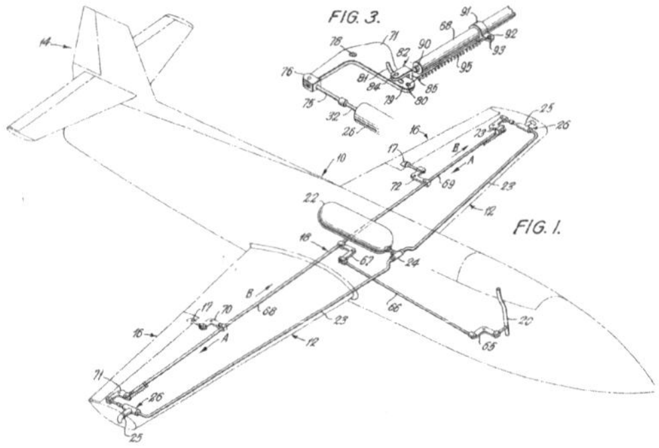

FIG. 3. FIG. 1.

June 2, 1964 McDONALD SILL 3,135,703

THRUST CHAMBER CATALYST STRUCTURE

Filed Sept. 21, 1959

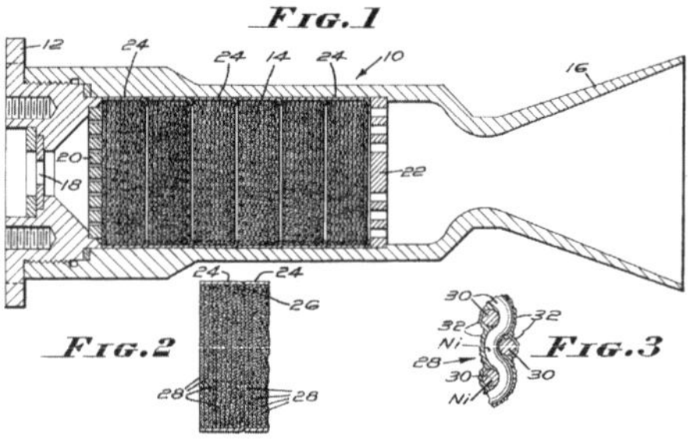

FIG. 1 FIG. 2 FIG. 3

Feb. 13, 1962 W. F. MOORE 3,021,095

PROPULSION UNIT

Filed June 10, 1960 3 Sheets-Sheet 1

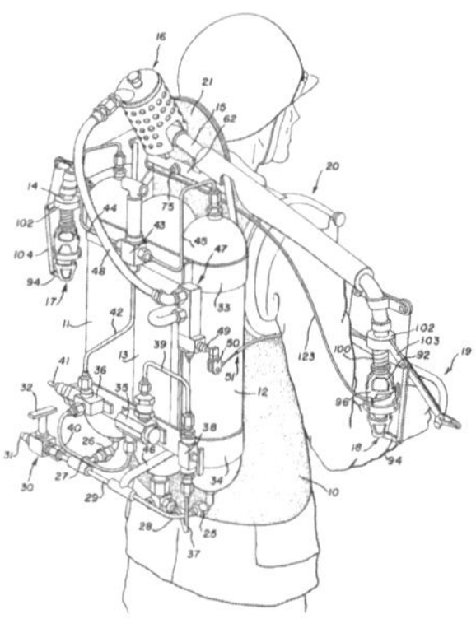

FIG. 1

Dec. 17, 1963 R. E. FLEXMAN 3,114,486

PACK CARRIER

Filed Sept. 22, 1961 2 Sheets-Sheet 1

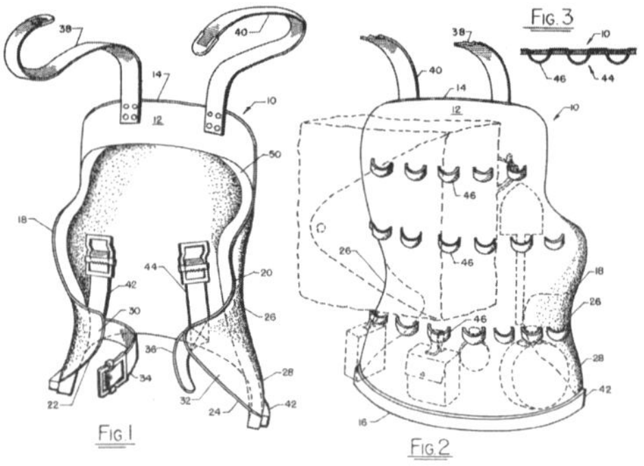

FIG. 1 FIG. 2 FIG. 3

Sept. 22, 1964 W. F. MOORE 3,149,798

INDIVIDUAL FLIGHT DEVICE

Filed Nov. 3, 1961 2 Sheets-Sheet 2

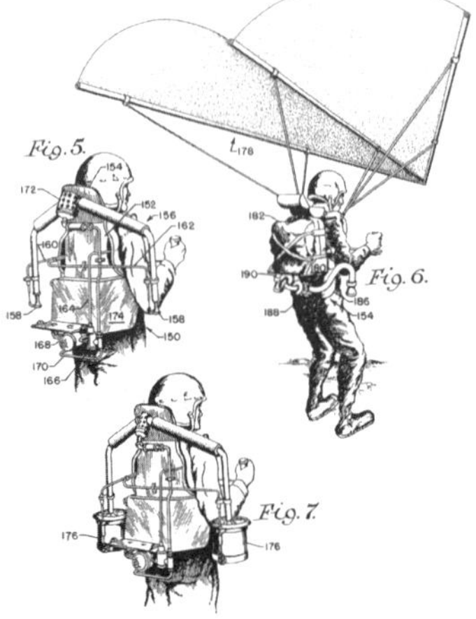

Fig. 5. Fig. 6. Fig. 7.

Dec. 4, 1962 W. F. MOORE 3,066,887

SPACE BELT

Filed May 9, 1960 3 Sheets-Sheet 1

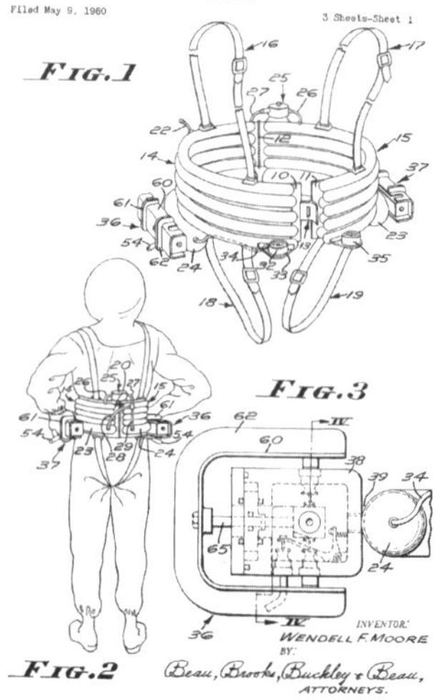

FIG. 1 FIG. 2 FIG. 3

INVENTOR
WENDELL F. MOORE
BY
Beau, Brooks, Buckley & Beau,
ATTORNEYS.

July 5, 1966 R. C. DEEGAN 3,258,959

THRUST MEASURING SYSTEMS

Filed Oct. 14, 1963 5 Sheets-Sheet 1

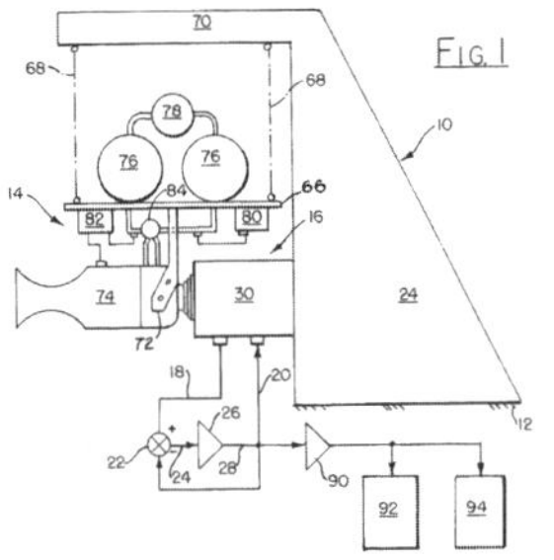

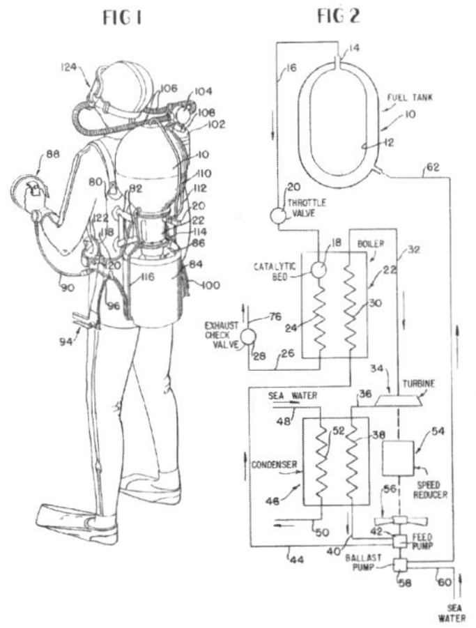

June 22, 1965 A. M. GORBATY 3,190,069

SPACE VEHICLE CONTROL SYSTEM

Filed Jan. 28, 1963 2 Sheets-Sheet 1

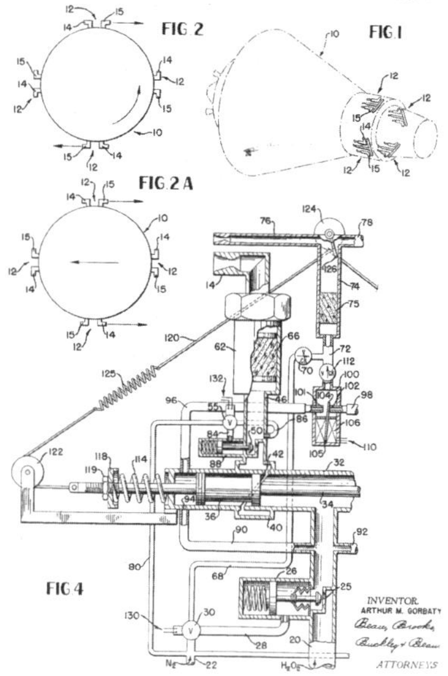

Sept. 22, 1964 J. K. HULBERT 3,149,799

INDIVIDUAL PROPULSION

Filed Sept. 19, 1963 5 Sheets-Sheet 1

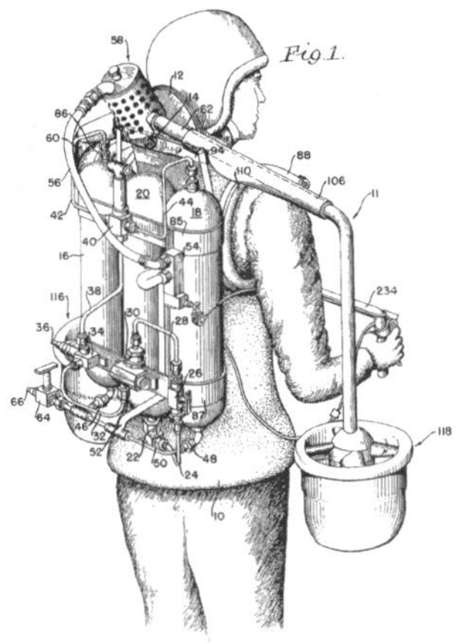

May 21, 1963 J. K. HULBERT 3,090,345

HYDRODYNAMIC PROPULSION SYSTEM

Filed Feb. 19, 1962 3 Sheets-Sheet 1

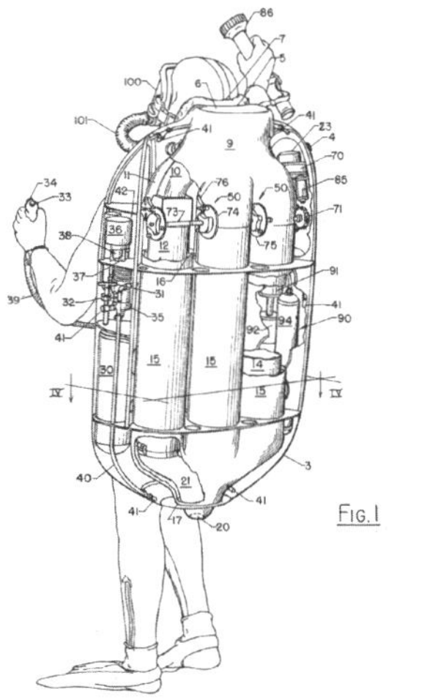

Dec. 27, 1966 J. K. HULBERT ET AL 3,293,851

UNDERWATER PROPULSION DEVICES

Original Filed March 29, 1963 3 Sheets-Sheet 1

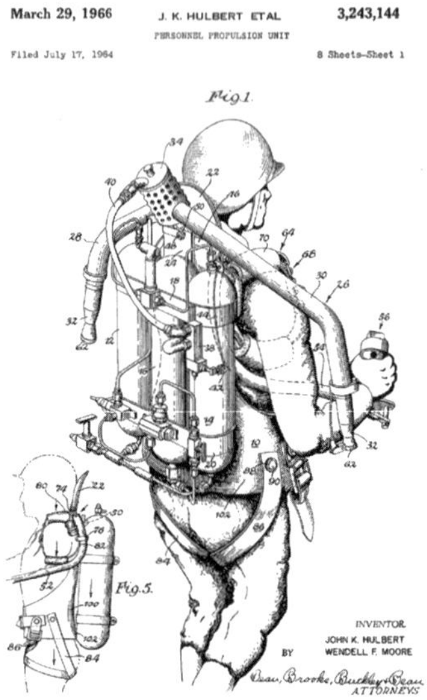

March 29, 1966 J. K. HULBERT ETAL 3,243,144

PERSONNEL PROPULSION UNIT

Filed July 17, 1964 8 Sheets-Sheet 1

Fig.1.

Fig.5.

INVENTOR.
JOHN K. HULBERT
WENDELL F. MOORE
BY
Dean, Brooks, Buckley & Bean
ATTORNEYS

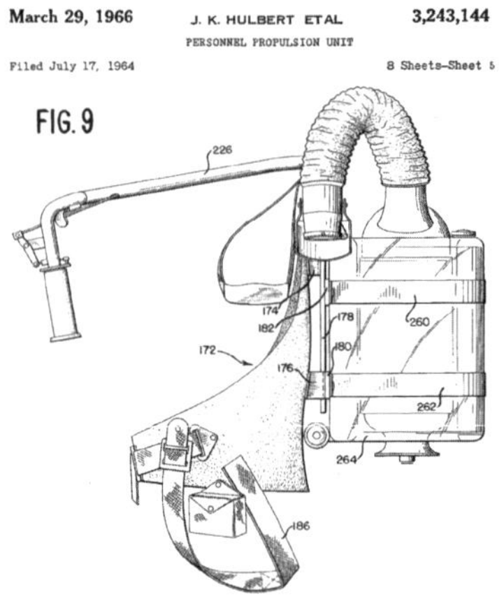

March 29, 1966 J. K. HULBERT ETAL 3,243,144

PERSONNEL PROPULSION UNIT

Filed July 17, 1964 8 Sheets-Sheet 5

FIG. 9

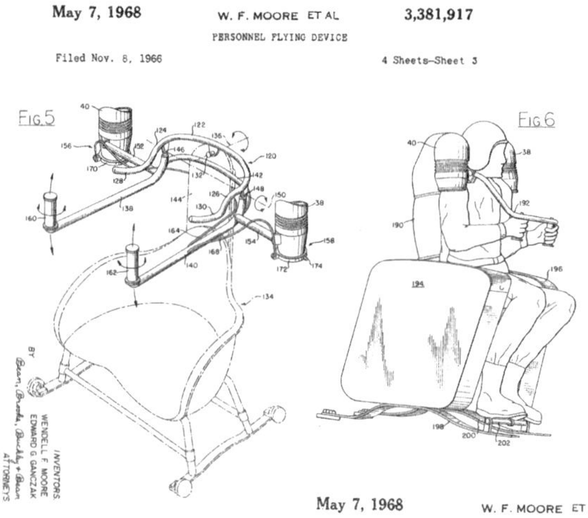

May 7, 1968 W. F. MOORE ET AL 3,381,917

PERSONNEL FLYING DEVICE

Filed Nov. 8, 1966 4 Sheets-Sheet 3

FIG 5

FIG 6

INVENTORS
WENDELL F. MOORE
EDWARD G. GANCZAK
BY
Dean, Brooks, Buckley & Bean
ATTORNEYS

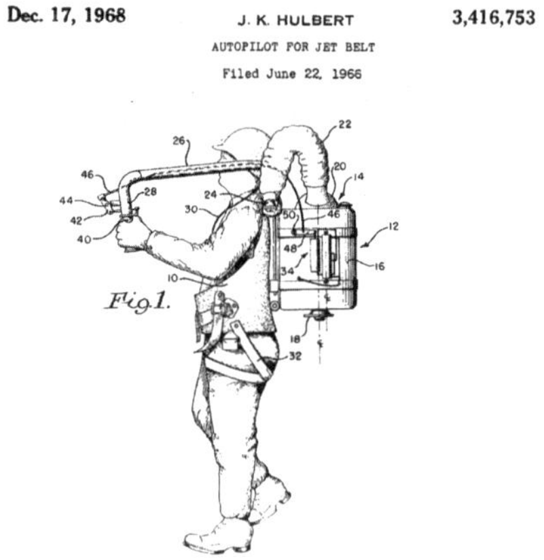

Dec. 17, 1968 J. K. HULBERT 3,416,753

AUTOPILOT FOR JET BELT

Filed June 22, 1966

Fig.1.

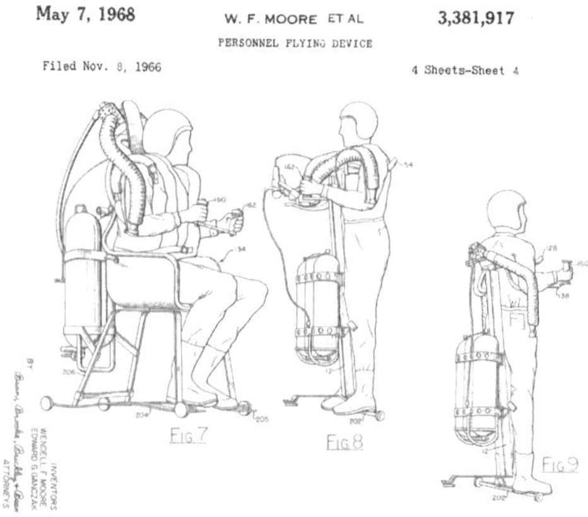

May 7, 1968 W. F. MOORE ET AL 3,381,917

PERSONNEL FLYING DEVICE

Filed Nov. 8, 1966 4 Sheets-Sheet 4

FIG 7

FIG 8

FIG 9

INVENTORS
WENDELL F. MOORE
EDWARD G. GANCZAK
BY
Dean, Brooks, Buckley & Bean
ATTORNEYS

Apogee Books Space Series

#	TITLE	ISBN	Bonus	US$	UK£	Can$
1	Apollo 8 NASA Mission Reports	978-1-896522-66-1	C	$18.95	£13.95	$25.95
2	Apollo 9 NASA Mission Reports	978-1-896522-51-7	C, w/o	$16.95	£12.95	$22.95
3	Friendship 7 NASA Mission Reports	978-1-896522-60-9	C	$18.95	£13.95	$25.95
4	Apollo 10 NASA Mission Reports	978-1-896522-68-5	C	$18.95	£13.95	$25.95
5	Apollo 11 NASA Mission Reports, Vol 1	978-1-896522-53-1	C	$18.95	£13.95	$25.95
6	Apollo 11 NASA Mission Reports, Vol 2	978-1-896522-49-4	C	$15.95	£10.95	$20.95
7	Apollo 12 NASA Mission Reports	978-1-896522-54-8	C	$18.95	£13.95	$25.95
8	Gemini 6 NASA Mission Reports	978-1-896522-61-6	C	$18.95	£13.95	$25.95
9	Apollo 13 NASA Mission Reports	978-1-896522-55-5	C	$18.95	£13.95	$25.95
10	Mars NASA Mission Reports	978-1-896522-62-3	C	$23.95	£18.95	$31.95
11	Apollo 7 NASA Mission Reports	978-1-896522-64-7	C, w/o	$18.95	£13.95	$25.95
12	The High Frontier	978-1-896522-67-8	C	$21.95	£17.95	$28.95
13	X-15 NASA Mission Reports	978-1-896522-65-4	C, w/o	$23.95	£18.95	$31.95
14	Apollo 14 NASA Mission Reports	978-1-896522-56-2	C	$18.95	£15.95	$25.95
15	Freedom 7 NASA Mission Reports	978-1-896522-80-7	C, w/o	$18.95	£15.95	$25.95
16	Shuttle STS 1-5 NASA Mission Reports	978-1-896522-69-2	C	$23.95	£18.95	$31.95
17	Rocket & Space Corporation Energia	978-1-896522-81-4		$21.95	£16.95	$28.95
18	Apollo 15 NASA Mission Reports	978-1-896522-57-9	C, w/o	$19.95	£15.95	$27.95
19	Arrows to the Moon	978-1-896522-83-8		$21.95	£17.95	$28.95
20	The Unbroken Chain	978-1-896522-84-5	C	$29.95	£24.95	$39.95
21	Gemini 7 NASA Mission Reports	978-1-896522-82-1	C	$19.95	£15.95	$26.95
22	Apollo 11 NASA Mission Reports, Vol. 3	978-1-896522-85-2	D, w/o	$27.95	£19.95	$37.95
23	Apollo 16 NASA Mission Reports	978-1-896522-58-6	C	$19.95	£15.95	$27.95
24	Creating Space	978-1-896522-86-9		$30.95	£24.95	$39.95
25	Women Astronauts	978-1-896522-87-6	C	$23.95	£18.95	$31.95
26	On To Mars	978-1-896522-90-6	C, w/o	$21.95	£16.95	$29.95
27	The Conquest of Space	978-1-896522-92-0		$23.95	£19.95	$32.95
28	Lost Spacecraft	978-1-896522-88-3		$30.95	£24.95	$39.95
29	Apollo 17 NASA Mission Reports	978-1-896522-59-3	C	$19.95	£15.95	$27.95
30	Virtual Apollo	978-1-896522-94-4		$24.95	£15.95	$27.95
31	Apollo EECOM	978-1-896522-96-8	C, w/o	$29.95	£23.95	$37.95
32	A Vision of Future Space Transportation	978-1-896522-93-7	C	$27.95	£21.95	$35.95
33	Space Trivia	978-1-896522-98-2		$19.95	£14.95	$26.95
34	Interstellar Spacecraft & Multi	978-1-896522-99-0		$24.95	£18.95	$30.95
35	Dyna-Soar	978-1-896522-95-1	D	$32.95	£23.95	$42.95
36	Rocket Team	978-1-894959-00-1	D	$34.95	£24.95	$44.95
37	Sigma 7 Mission Reports	978-1-894959-01-8	C	$19.95	£15.95	$27.95
38	Women of Space	978-1-894959-03-2	C	$22.95	£17.95	$30.95
39	Columbia Accident Report	978-1-894959-06-3	C, w/o	$25.95	£19.95	$33.95
40	Gemini 12 Mission Reports	978-1-894959-04-9	C	$19.95	£15.95	$27.95
41	Simple Universe	978-1-894959-11-7		$21.95	£16.95	$29.95
42	New Moon Rising	978-1-894959-12-4	D	$33.95	£23.95	$44.95
43	Moonrush	978-1-894959-10-0		$24.95	£17.95	$30.95
44	Mars Mission Reports, Vol 2	978-1-894959-05-6	D	$28.95	£20.95	$38.95
45	Rocket Science	978-1-894959-09-4		$20.95	£15.95	$28.95
46	How NASA Learned to Fly	978-1-894959-07-0		$25.95	£18.95	$35.95
47	Virtual LM	978-1-894959-14-8	C	$29.95	£22.95	$42.95
48	Deep Space Mission Reports	978-1-894959-15-5	D	$34.95	£22.95	$44.95

Bonus legend: C = CD-ROM, D = DVD, w/o = available from our web site only.

http://www.apogeebooks.com

Apogee Books Space Series

#	TITLE	ISBN	Bonus	US$	UK£	Can$
49	Space Tourism	978-1-894959-08-7		$20.95	£15.95	$28.95
50	Apollo 12 Mission Reports, Vol 2	978-1-894959-16-2	D	$24.95	£15.95	$31.95
51	Atlas	978-1-894959-18-6		$29.95	£16.95	$37.95
52	Reflections from Orbit	978-1-894959-22-3		$23.95	£16.95	$30.95
53	Real Space Cowboys	978-1-894959-21-6	D	$29.95	£17.95	$36.95
54	Saturn	978-1-894959-19-3	D	$27.95	£18.95	$35.95
55	On To Mars 2	978-1-894959-30-8	C	$22.95	£14.95	$29.95
56	Getting Off the Planet	978-1-894959-20-9		$18.95	£12.95	$23.95
57	Return to the Moon	978-1-894959-32-2		$22.95	£15.95	$28.95
58	Beyond Earth	978-1-894959-41-4		$27.95	£18.95	$36.95
59	ISScapades	978-1-894959-59-9		$23.95	£12.95	$27.95
60	Astronautics Book 1	978-1-894959-63-6		$23.95	£12.95	$27.95
61	Go For Launch	978-1-894959-43-8		$29.95	£18.95	$34.95
62	Reference Guide to ISS	978-1-894959-34-6		$21.95	£10.95	$23.95
63	Around World 84 Days	978-1-894959-40-7		$31.95		
64	Spaceships	978-1-894959-50-6		$19.95	£10.95	$22.95
65	Voice of Wernher von Braun	978-1-894959-64-3		$22.95	£12.95	$24.95
66	Surveyor Mission Reports	978-1-894959-65-0		$17.95	£9.95	$21.95
67	Astronautics Book 2	978-1-894959-66-7		$25.95	£13.95	$26.95
68	End of the Solar System	978-1-894959-68-1		$26.95		
69	Lunar Exploration Scrapbook	978-1-894959-69-8		$36.95	£16.95	$36.95
70	Apollo Advanced Lunar Exploration Plan	978-1-894959-80-3	w/o	$16.96		
71	Canada's Fifty Years in Space	978-1-894959-72-8		$26.95		
72	Floating to Space	978-1-894959-73-5	D	$27.95		
73	The Astronaut & The Fireman	Cancelled				
74	Saturn I/IB	978-1-894959-85-8	D	$26.95		
75	Cold War Tech War	978-1-894959-77-3		$29.95		
76	Krafft Ehricke's Extraterrestrial Imperative	978-1-894959-91-9		$27.95		
77	License to Orbit	978-1-894959-98-8		$27.95		
78	The Nuclear Rocket	978-1-894959-99-5		$21.95		
79	Apollo 17 NASA Mission Reports Vol 2	978-1-926592-02-2	D	$26.95		
80	Lightcraft Flight Handbook	978-1-926592-03-9		$29.95		
81	Energy Crisis: Solution From Space	978-1-926592-06-0		$24.95		
82	Rocket Belt Pilot's Manual	978-1-926592-05-3		$22.95		
83	Selling Peace	978-1-926592-08-4		$TBA		
84	The Farthest Shore	978-1-926592-07-7		$TBA		

Apogee Books Pocket Space Guides

#	TITLE	ISBN	US$	UK£	Can$
1	Apollo 11	978-1-894959-27-8	$9.95	£6.95	$12.95
2	Mars	978-1-894959-26-1	$9.95	£6.95	$12.95
3	Deep Space	978-1-894959-29-2	$9.95	£6.95	$12.95
4	Launch Vehicles	978-1-894959-28-5	$9.95	£6.95	$12.95
5	Apollo Test Program	978-1-894959-36-0	$9.95	£6.95	$10.95
6	Apollo Exploring	978-1-894959-37-7	$9.95	£6.95	$10.95
7	Hubble Telescope	978-1-894959-38-4	$9.95	£6.95	$10.95
8	Russian Spacecraft	978-1-894959-39-1	$9.95	£6.95	$10.95
9	Project Constellation	978-1-894959-49-0	$9.95	£6.95	$11.95
10	Space Shuttle	978-1-894959-52-0	$9.95	£6.95	$11.95
11	Project Mercury	978-1-894959-53-7	$9.95	£6.95	$11.95
12	Project Gemini	978-1-894959-54-4	$9.95	£6.95	$11.95